CONCURRENT ENGINEERING

CONCURRENT ENGINEERING

Concepts, implementation and practice

Edited by

Chanan S. Syan

Department of Mechanical Engineering and Manufacturing Systems, University of Northumbria at Newcastle, Newcastle upon Tyne, UK

and

Unny Menon

Department of Industrial Engineering, California Polytechnic State University, San Luis Obispo, California, USA

SPRINGER-SCIENCE+BUSINESS MEDIA, B.V.

First edition 1994

© 1994 Springer Science+Business Media Dordrecht
Originally published by Chapman & Hall in 1994
Softcover reprint of the hardcover 1st edition 1994

Typeset in 10/12 pt Palatino by Type Study, Scarborough

ISBN 978-94-010-4566-7   ISBN 978-94-011-1298-7 (eBook)
DOI 10.1007/978-94-011-1298-7

A catalogue for this book is available from the British Library

Library of Congress Catalog Card Number: 94–71062

∞ Printed on permanent acid-free text paper, manufactured in accordance with ANSI/NISO Z39.48–1992 and ANSI/NISO Z39.48–1984 (Permanence of Paper)

To my wife Jasbir, my daughter Gupinder, my son Dalvir and the Syan family, for their support and who have put up with constant neglect for the last two years during the compilation of this book.

C. S. Syan

To Pauline, for all the support she gave

U. Menon

Contents

PART THREE **Product Design, Support and Management Tools for Concurrent Engineering**

Contributors

Mr J. D. A. Anderson,
European Regional Director,
ICAD Engineering Automation Ltd.,
Viscount Centre II, Milburn Hill Road,
University of Warwick Science Park,
Coventry, CV4 7HS,
UK.

Professor J. V. Chelsom,
Professor of Engineering Management,
Department of System Science,
City University, Northampton Square,
London, EC1V 0HB,
UK.

Dr P. M. Dickens,
Department of Manufacturing Engineering and Operations
    Management,
University of Nottingham, University Park,
Nottingham, NG7 2RD,
UK.

Mr A. Grazebrook,
I-Logix UK Ltd.,
Bumpers Way, Chippenham,
Wiltshire, SN14 6RA,
UK.

Mr J. Z. Gu,
Group for Intelligent Systems in Design and Manufacture,
Department of Industrial Engineering,
North Carolina State University,
Raleigh, North Carolina 27695–7906,
USA.

Mr S. C. Hitchins,
Product Marketing Manager,
Computervision Ltd., Argent Court,
Sir William Lyons Road,
Coventry, CV4 7EZ,
UK.

Professor U. Menon, PHD, FIMECHE,
Industrial Engineering Department,
California Polytechnic State University,
San Luis Obispo, CA 93407,
USA.

Professor C. O'Brien,
Department of Manufacturing Engineering and Operations
    Management,
University of Nottingham, University Park,
Nottingham, NG7 2RD,
UK.

Professor P. J. O'Grady,
Group for Intelligent Systems in Design and Manufacture,
Department of Industrial Engineering,
North Carolina State University,
Raleigh, North Carolina 27695–7906,
USA.

Dr K. S. Pawar,
Department of Manufacturing Engineering and Operations
    Management,
University of Nottingham, University Park,
Nottingham, NG7 2RD,
UK.

Ms S. Schedler,
Logistics IT Strategy Programme Manager,
Integraph (UK) Limited,
Delta Business Park, Great Eastern Way,
Swindon, Wiltshire, SN5 7XP,
UK.

Mr S. J. Smith,
Department of Manufacturing Engineering and Operations
    Management,
University of Nottingham, University Park,
Nottingham, NG7 2RD,
UK.

Professor K. G. Swift, BSC, MSC, PHD, CENG, MIMECHE, MIEE,
Department of Engineering Design and Manufacture,
University of Hull, Cottingham Road,
HU6 7RX,
UK.

Professor C. S. Syan, CENG, BENG, PHD, MIMECHE, MIEE, MBCS,
Department of Mechanical Engineering and Manufacturing Systems,
University of Northumbria at Newcastle,
Newcastle upon Tyne, NE1 8ST,
UK.

Dr R. E. Young,
Group for Intelligent Systems in Design and Manufacture,
Department of Industrial Engineering,
North Carolina State University, Raleigh,
North Carolina 27695–7906.
USA.

Acknowledgements

Whilst preparing this book, many friends gave valuable support and help. We will not attempt to compile a list as we do not want to omit anybody. Instead, we would like to take this opportunity to thank everyone who gave advice, made constructive comments and contributed to the compilation of this work.

It is not intended for this text to be read from cover to cover, but, referred to selectively, as a handbook for specific areas of CE and its implementation. We wish all companies committed to success through CE, product development teams and students of CE every success in their quest. Good luck.

C. S. Syan
University of Northumbria at Newcastle
UK

U. Menon
California Polytechnic State University
USA

Introduction

BACKGROUND

There is an increasing awareness that 'time to market' is the key competitive issue in the manufacturing industry today. The global markets are demanding products that are well designed, are of high quality and are at low prices with ever decreasing lead times. Hence manufacturers are forced to utilize the best methods of technology with efficient control and management accompanied by suitably enabling organizational structures.

Concurrent engineering (CE) is widely seen to be the methodology that can help satisfy these strenuous demands and keep the profitability and viability of product developers, manufacturers and suppliers high. There have been many reported successes of CE in practice. Rover were able to launch Land Rover Discovery in 18 months as compared with 48–63 months for similar products in Europe. Because of its early introduction to the market, it became the best selling product in its class. AT&T report part counts down to one ninth of their previous levels and quality improvements of one hundred times (in surface defects) for VLSI (very large scale integration) circuits as a result of using the CE approach.

WHO SHOULD READ THIS TEXT?

This book will aim to provide a sound basis for the very diverse subject known as concurrent engineering. Concurrent engineering is recognized by an increasingly large proportion of the manufacturing industry as a necessity in order to compete in today's markets. This recognition has created the demand for information, awareness and training in good concurrent engineering practice.

This text will introduce the subject area, identify the major elements, tools and techniques and procedure in a single source. The important areas of this approach will be explained and discussed by experts in these fields. There will also be a comprehensive guide to important further references, sources of information and centres of excellence in Europe, the USA and the Far East.

The objectives of this publication are to:

1. provide a comprehensive single source guide to the subject;
2. act as the main course text for the successful short courses in concurrent engineering industrial courses run by the book editors;
3. be a reference source for researchers and final year students of engineering courses and advanced graduate courses;
4. be a reference for designers and manufacturing engineers in electronic, mechanical, aerospace, automotive, food and other engineering disciplines.

The major drive for this book is 2. and 3. above, as the editors run industrial graduate and undergraduate courses in the concurrent engineering field. This book will become the recommended text for all these courses.

ABOUT THIS BOOK

The domain of CE is very wide and includes elements such as management of change, the team approach, the design process and its management, marketing, puchasing and procurement, manufacturing, distribution and support. This text is a culmination of the editors' work in training, working with industry in implementing CE and research and development effort over a number of years. All the chapters are written by practitioners and experts in their fields. The majority of the content of this book is based on an established three-day course on CE run in the UK by the editors for industry.

The organization of the book

The contents of this book fall into three distinct parts. These are:

Part One
Chapters 1 to 4 introduce, overview and define CE. Case studies are presented for deeper understanding of the practice and important issues are highlighted. The organizational and management approaches to CE practice are also introduced.

Part Two
Chapters 5 to 9 look at the essential tools and techniques for successful and efficient CE practice. They introduce the state of the art in quality function deployment, rapid prototyping and design for assembly and manufacture.

Part Three
Chapters 10 to 13 introduce currently available CE support tools. This section is not meant to be exhaustive, as that would be an impossible task for a single text on the subject. It aims to show a range of products and their features and capabilities. Where possible, the system capabilities are illustrated via user case studies.

Although the text is not physically partitioned into three distinct parts, the chapters are designed to fall into the above categories.

Chapters and authors

This section introduces and summarizes the contents of each chapter of this book. It also gives some information about the authors and their affiliations.

Chapter 1 introduces CE. The basic principles, approaches, benefits and CE practice are discussed. The general rules derived from a wide study of this topic are given to help those embarking upon the route to CE practice. The chapter is written by Dr Chanan S. Syan from the University of Sheffield, UK. He has been involved with design for manufacture and CE for more than ten years. He regularly runs CE courses for, and works closely with, industry, and has active research interests in related areas.

Chapter 2 gives the experiences of the Ford Motor Company with CE. Ford was one of the first key players in this field. The two case studies give a comprehensive insight into two projects undertaken by Ford Europe in the early eighties. Professor John Chelsom is the author. He is from City University, UK. He is well placed to give comments and advice on these projects and was the marketing director of Ford Europe at the time when the reported projects were being run.

Chapter 3 looks at the organization structures that have been found to lend themselves to the practice of CE. The management issues of team building, motivation, rewards, and so on are introduced. The author, Dr Kulwant S. Pawar, is a lecturer at the University of Nottingham, UK. He also regularly lectures on the CE courses run by the editors.

Chapter 4 concerns design maturity. Inevitably in product development there are several sequential activities throughout the phases. There is a difficulty in assessing when a design stage has been sufficiently defined so that the next phase can be passed on to. Conversely, the design may not have been sufficiently defined in the proceeding phase. In both cases excessive dialogue, misinterpretation and re-design can result, leading to costly time delays. This chapter describes the current state of the art in this field. The co-authors of this work, Professor Chris O'Brien and Steve Smith, are from University of Nottingham, UK. They are currently studying this problem in the electronics industry.

Chapter 5 examines Quality Function Deployment. This technique can help interpret customer needs into design, manufacturing, packaging and maintenance issues. Quality Function Deployment also helps build team spirit and group work practices. The co-authors for this chapter are Professor Unny Menon from the Cal Poly State University, USA and Professor Peter J. O'Grady, Jiao Zhong Gu and Dr. Robert E. Young from North Carolina State University, USA.

Chapter 6 is intended to give a thorough understanding of the issues and methods used in this area. The chapter contains a review, analysis and description of the various elements of design for manufacture (DFM). The co-authors are Dr Chanan S. Syan and Professor Ken G. Swift. Professor Swift is from the University of Hull, UK. The authors have extensive experience in DFM/DFA areas, both in undertaking research and working with industry.

Chapter 7 deals with design for assembly (DFA) and presents the state of the art in this important area. The methods and the commercially available tools are introduced and described. The co-authors are, as for Chapter 6, Dr Chanan S. Syan and Professor K. G. Swift.

Chapter 8 looks at the rapid prototyping process of physical parts. The ability for the design engineer and the CE team to get a prototype made quickly from the design model is an extremely useful facility. This chapter introduces the techniques for physical component prototyping systems. The author is Dr P. M. Dickens from University of Nottingham.

Chapter 9 describes the approaches to prototyping of computer, electronic and hybrid systems. These techniques enable rapid software emulations of complex systems without the necessity for large investments in traditional product development approaches. The author is Mr Alvery Grazebrook from I-Logix (UK), who market such a prototyping system worldwide.

Chapter 10 concerns software tools for the product development process. Computervision's CADDS products, including assembly mock-up, information, control and engineering data management (EDM) systems, are described. The capabilities of these packages are illustrated by example cases. The author is Steve Hitchins from Computervision Ltd., UK.

Chapter 11 looks at the role of knowledge-based systems (KBS) in CE. The features, suitably and current uses of KBSs are discussed and the future possible applications are outlined. The author is Jeremy D. A. Anderson of ICAD Engineering Automation, UK. ICAD stands for intelligent CAD (feature-based) system which utilizes KBS approaches.

Chapter 12 introduces the Integraph CAD/CAM system as well as the supporting products for CE. These include the family of technical information management systems (TIM). As with other chapters in this section, the work includes industrial applications to aid clarity and understanding. The author is Ms Sue Schedler of Integraph (UK) Ltd.

Finally, Chapter 13 is a brief summary of the strategies for concurrent engineering and lists some useful sources of further information and readings on the subject. The authors are Dr Chanan S. Syan and Professor John Chelsom.

PART ONE

Concurrent Engineering: Concepts, Definitions and Issues

CHAPTER 1

Introduction to concurrent engineering

C. S. Syan

1.1 BACKGROUND

Throughout the world for the past two decades, all engineering companies have faced similar challenges. These are ever more demanding customers, rapid technological change, environmental issues, competitive pressures on quality and cost, and shorter time to market with additional new product features. This is all happening with the majority of the Western world's common economic background of slow growth, excess capacity, increasing legislation compliance, demographic changes, market complexity and increasing globalization of industries.

In many cases the pace was set by the Japanese, who progressively made inroads in North America and Europe and in some cases dominated chosen markets. The list of these chosen markets became longer year by year. Western companies were slow to recognize the basis of Japanese success, but eventually responded with a whole string of actions including CAD/CAE/CAM/CIM, robotics, automation, value analysis, quality programmes, information technology and so on. They sought to offset a perceived weakness – their workforces – by building on their apparent strength – technology, particularly computer-based technology.

This expensive technology was largely ineffective, because the new tools were used with existing structures, practices and attitudes. Products continued to arrive in the market place at unsatisfactory quality levels, and often too late to achieve sales and profit objectives. Their efforts were also undermined by short-term successes during market booms. There were also brief respites as a result of the appreciating yen, but the Japanese overcame this by aggressive cost reduction programmes and by investments in production capacity outside Japan – sometimes in countries with lower labour costs, sometimes in the key sales territories.

In the 1980s, companies started to feel the influence of large multinational organizations on the markets, increased product complexities and new developments in innovative technologies. This directly affected the organization's ability to develop and introduce new products to the market. This was especially true for the electronics industry, where product lives were reducing significantly.

To compete successfully, companies have to continuously keep reducing development times and sustain improvements in their products and their quality. The need for better quality and shorter product development lead times is now widely acknowledged and the realization that the concurrent engineering approach offers the best way of achieving these objectives has also become a necessary company strategy.

1.2 SEQUENTIAL ENGINEERING

In order to understand CE, it is useful to describe the traditional introduction and product development practice of the majority of the Western world's manufacturing companies. Typically, in a manufacturing organization, marketing identifies the need for new products, price ranges and their expected performance from customers or potential consumers. Design and engineering receive loose specifications and commonly work alone developing the technical requirements (e.g. materials and size) and final design detail as well as the associated documentation such as drawings and bills of materials etc.

As design is carried out in relative isolation, manufacturing, test, quality and service functions only see the design in an almost complete state. As the process is sequential in progression, each stage of product development following completion of the previous stage, it is commonly known as sequential engineering. Since the design for any new product arrives in the manufacturing department with about as much warning and involvement as if it had been thrown over the factory wall, it is also commonly known as 'over the wall engineering'. Figure 1.1 illustrates the sequential process of new product development, where each design stage starts only when the previous one is completed. This type of approach is also known by many other names, including serial engineering, time-phased engineering, and the chimney method.

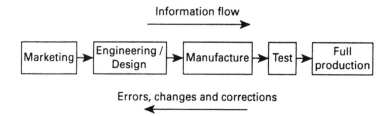

Fig. 1.1 The sequential engineering process.

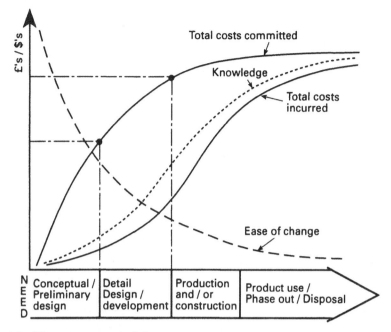

Fig. 1.2 Time versus cost of changes.

In this sequential method of operation, a change required in a later stage will cause delay and additional costs in the upstream stages. Additionally, the subsequent stages will be delayed until the current stage has been completed. This approach encourages a large number of modifications and alterations in the later stages of the product development phase, when it is more expensive and difficult, as shown in Figure 1.2. In many cases investment in tooling and equipment is usually committed and the product launch date may already be fixed.

There are many weaknesses of the over the wall engineering approach. In summary, they include:

- insufficient product specification, leading to an excessive amount of modifications;
- little attention to manufacturability issues of the product at the design stage;
- the estimated costings are usually degrees of magnitude in error, due mainly to the uncontrolled late design change costs. This leads to a lack of confidence in the estimated costs of projects;
- the likelihood of late changes usually leads to expensive changes to tooling and other equipment.

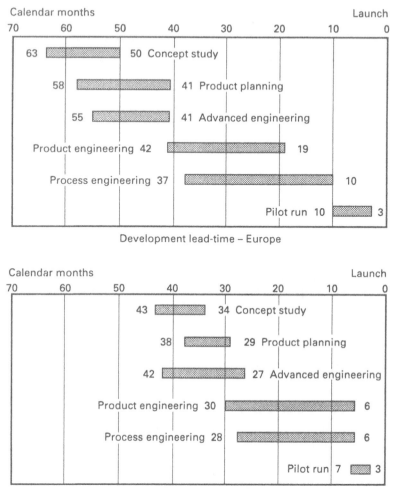

Fig. 1.3 Comparison of European and Japanese development lead times for the automotive industry

1.3 CONCURRENT ENGINEERING

Concurrent engineering is also known by many other names, including simultaneous engineering, concurrent design, life-cycle engineering, integrated product development and team design. Because CE is a dramatic departure from the past practice, there is an immediate need for a new design environment and technology requiring extensive interdisciplinary co-operation and integration of such fields as design engineering, manufacturing, material management and marketing. However, the concept of CE is not new. It has been practised by successful manufacturing managers, but no one has paid much attention to applying it in a systematic way. Japanese industry has practised CE, without using its name, for some time. This is clearly illustrated by the studies done in the automotive industry, comparing the time to market of Japanese and European manufacturers. Figure 1.3 shows that for 12 projects studied, typical Japanese companies could develop and introduce a new car to market in 43 months against 63 months for the 11 projects studied in Europe [1].

1.3.1 CE definition and requirements

A very significant event in the history of CE took place in the USA in 1982. The Defense Advanced Research Projects Agency (DARPA) started a study into improving concurrency in the design process. In 1986, the Institute for Defense Analysis (IDA) Report R-338 coined the term 'concurrent engineering' to explain the systematic method of product and process design, as well as other support processes and services. The IDA report also gave a definition of concurrent engineering which is now widely accepted. This definition of CE is as follows:

> Concurrent engineering is a systematic approach to the integrated, concurrent design of products and their related processes, including manufacture and support. This approach is intended to cause the developers, from the outset, to consider all elements of the product life cycle from concept through disposal, including quality, cost, schedule, and user requirements.

CE provides a systematic and integrated approach to introduction and design of products. The subsets of CE include design for manufacture, design for assembly, design for maintainability, design for disposal and so on. Figure 1.4 illustrates the concurrent engineering approach graphically. Functions such as design and engineering are integrated in terms of continuous and complete information exchanges. As the commencement of each distinct stage is not dependent upon full completion of the preceding stage, overlapping activities can take place, leading to concurrency in product development.

Effective CE practice requires good communications between disparate functions associated with the product life-cycle. The information must have common ownership, be shared freely and must be easily and freely accessible. As information is seen to be power in functionally organized traditional companies, this suggests more open organizational structures such as matrix management and team work. CE is therefore the integration of all company resources needed for product development, including people, tools and resources, and information.

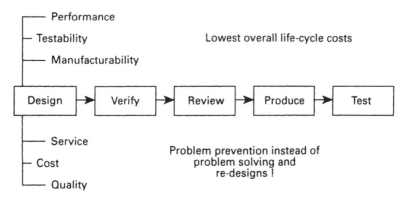

Fig. 1.4 The concurrent engineering process.

1.3.2 Objectives of CE

Studies considering the costs associated to a product during its entire life-cycle have demonstrated that from 60 to 95% of these costs are determined during the design phase. Therefore, it is during the design that the best savings can be achieved. Moreover, the earlier the improvements are made the greater is the cost reduction, as shown in Figure 1.5. The purpose of concurrent engineering is to ensure that the decisions taken during the design of a product result in a minimum overall cost during its life-cycle. In other words, this means that all activities must start as soon as possible, to induce working in parallel, which additionally shortens the overall product development process.

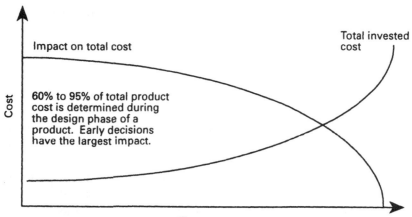

Fig. 1.5 The importance of the design on the product life-cycle cost.

The main objectives of concurrent engineering may be summarized as follows:

- decreased product development lead-time;
- improved profitability;
- greater competitiveness;
- greater control of design and manufacturing costs;
- close integration between departments;
- enhanced reputation of the company and its products;
- improved product quality;
- promotion of team spirit.

1.3.3 Why use concurrent engineering?

The level of competition in all markets, including engineering products, is globally increasing. Reasons for this are complex, but the main contributors are use of new technology, larger number of organizations in the same markets and wider appreciation and use of continuous process improvements. Goldhar *et al.* [2] describe a set of dominant trends in the business environment that have influenced the competitiveness of the companies since 1980. These are the product life-cycles of products shortening, the diversity, variety and complexity of products increasing and the customers becoming increasingly more sophisticated and demanding customized products more closely targeted to their needs. This has led to pressures on continuous product improvements, leading to ever increasing functionality and features.

Figure 1.6 shows the reducing trends in product lifetimes and the increasing development times. The consequences for a company would be disastrous if the development times are not reduced significantly whilst still maintaining quality and keeping the costs down. The increase

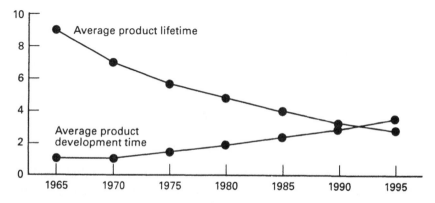

Fig. 1.6 Product development time versus the product's lifetime (for electronics products [7]).

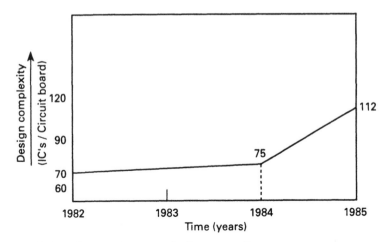

Fig. 1.7   Increasing product complexity versus time.

in product development times is due in part to the increasing complexity of the products. Figure 1.7 shows the trend of product lifetimes versus development times in the electronics industry. For companies to survive and remain competitive in world markets, they have to reduce the product development lead-times significantly.

Delays in bringing a product to market certainly result in greater losses of profit. Carter and Baker [4] introduced a simple method to measure the impact of delays in launching a product, as shown in Figures 1.8a and 1.8b. For example, considering a 12-month market window, a delay of two months in launching a new product will result in 24% loss in total lifetime revenue.

The profit therefore comes from early product launch, hence bringing products faster to the market, which allows the achievement of the optimum profitability. However, it is necessary to attack the causes of delay in the development process. As the products have a short lifetime, no time is allowed for the companies to correct design errors or re-engineer products to higher quality at low costs. Therefore it is essential to renounce a philosophy of 're-do until right', and introduce the philosophy of 'right first time'. Concurrent engineering is indispensable to companies that desire to remain competitive, improve their products and processes continuously and keep their development ahead of the competition.

1.3.4   Benefits of CE

The execution of the activities of the design in parallel leads to improvements in many areas such as communication, quality, production processes, cash flows and profitability. The reductions of time to market, which has strategic importance, allows companies to increase

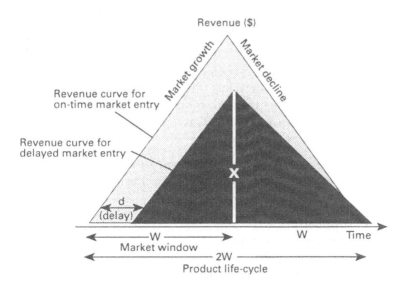

(a) Source : Logic Automation

$$\frac{d(3w-d)}{2w^2}$$

Months late	Revenue lost
1	12%
2	24%
3	34%
4	44%
5	54%

d = delay in months

(b) Source : Logic Automation

Fig. 1.8 Revenue loss due to delay to introduce a new product.

their market share and reduce design changes and design iterations. They are more easily manufacturable, serviceable and are of higher quality. Once released to manufacturing, production progresses quickly to full volume because the process is well defined, documented and controlled.

The remarkable performance achieved by world-class companies has been the best proof of the effectiveness of concurrent engineering. Their success has been recorded in books and articles, reporting striking improvements in terms of cycle times, cost reduction, product quality and reliability. Turino [5], for instance, reports that Boeing's Ballistic System Division achieved the following improvements.

- 16% to 46% in cost reduction in manufacturing.
- Engineering changes reduced from 15–20 to 1–2 drafts per drawing.
- Design analyses for the '-ilities' (e.g. design for manufacturability, etc.) cut from two weeks to less than one hour.
- Materials shortage reduced from 12% to 1%.
- Inspection costs cut by a factor of 3.

Business Week [6] describes the development of a new electronic cash register at NCR, highlighting the following benefits achieved:

- reduction in parts and assembly line;
- 65% fewer suppliers;
- 100% fewer screws or fasteners;
- 100% fewer assembly tools;
- 44% improvement in manufacturing costs;
- a trouble-free product introduction.

Other examples are: Rolls-Royce reduced the lead-time to develop a new aircraft engine by 30%; McDonnell Douglas reduced production costs by 40%; and ITT reduced their design cycle-time by 33% for its electronics counter measuring system. Many other cases which corroborate the benefits of adopting concurrent engineering are also reported by Turino [5], Ettlie [7] and in IEEE Spectrum [3].

It is possible to deduce some common factors from a number of successful cases reported in literature. It is widely quoted that when moving from the sequential to the concurrent approach, it is important that a company sets specific targets for itself. Lucas Automotive, for example, established the following targets when it moved into CE:

- 50% reduction in lead time to delivery;
- 50% reduction costs;
- zero defects;
- simplified procedures;
- design standardization.

1.4 SUPPORT FOR CE

CE can be practised without computer support or by using any formal techniques. The project team members can input their expertise and experience and achieve good results. This type of CE practice exists in very small companies who have very skilled and experienced people in their organizations. However, there is no guarantee of the variety of expertise being available in all organizations, hence it is desirable to provide support in various tasks of product development for improving performance.

There is also a large amount of information that needs to be communicated and this may be in many different forms such as drawings, data, electronic text, etc. For companies which have disparate sites, of about twenty or more employees, or are very large, this level of communication can become difficult to achieve. Hence networking, electronic data interchange (EDI) and computer assistance may be a pre-requisite to successful CE practice.

Most companies have automated support for various functions such as the use of computer-aided design and computer-aided manufacturing (CAD/CAM) systems in design, engineering data management systems and so on. There are also numerous computer support facilities for engineering and manufacturing as well as formal methods being used by many companies. It is essential that these facilities are integrated into the CE approach taken by an organization for optimal benefit.

There are four broad classes of support for CE activity, which are:

- process initiatives;
- computer-based support;
- formal techniques;
- data interchange methods.

1.4.1 Process initiatives

There are mainly two aspects of this type of support for CE. One is the team formation and operation, including its management and support. The other is the organization of structural and cultural change to accommodate and to enable the team approach to work effectively. There are other more specific support systems for the CE teams, which in the majority of cases are also available as computer-based systems.

(a) The team approach

Working in teams is not usual practice for most of the Western world's companies. In order to achieve effective team working, a major structural and organizational development process needs to take place. All the team members must work as part of a group with the same objectives, hence the selection process undertaken by the management must be precise. The team created should appreciate that their role is not of a committee, but actually to do what is required to successfully complete the project in hand. The CE practice demands multidisciplinary team approach. As a minimum, this task force should contain individuals from the departments shown in Figure 1.9, including the main suppliers and the customer.

CE task force composition

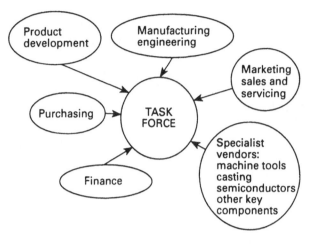

As a minimum the task force should have people from the
departments shown above, and the main vendors

Fig. 1.9   The minimum composition of the CE task force.

Hartley and Mortimer [1] suggest that in setting up a task force, there
are three basic options.

1. At the pre-concept stage a small team drawn from product design,
   manufacturing and finance may be sufficient. This needs to be
   expanded once the product moves from the concept to development
   stage.
2. A full workforce to take the product from the pre-concept state
   through to full production.
3. A task force which will take the product from the pre-concept stage
   and remain with the product as long as it is in production. Such a team
   is likely to vary in composition and size depending upon the stage of
   product development and manufacture.

Multidisciplinary teams must be given near full authority for the
decisions regarding the project, being responsible for the entire development of a new product. High authority also increases motivation and
commitment of the team members, establishing an environment of
common ownership and identity with the project. This activity requires
training of both the team members and the management in order to be
effective. There can be many problems and difficulties that need to be
sorted out, such as level of autonomy, authority and accountability. Also
issues such as team member selection, motivation and time management,
perhaps between more than one project, can be difficult problems in
practice to solve.

(b) Team duties

For a multidisciplinary team, effective communication amongst the members is extremely crucial for practising concurrent engineering successfully. The types of communication needed are between all the people involved and programs used such as CAD/CAM etc. in product development.

The activities that teams undertake are varied and very wide-ranging. The use of handbooks and data books is a common activity, as is using various analytical methods. Communications between various functional activities, such as design and engineering, and face-to-face meetings with other team members enable negotiations, compromise and decisions to be made, based upon the experience of the team members. Perhaps the most important activity that the product development team undertakes is the documentation and archiving of information and decisions made.

Co-location of people provides transparency in ideas and issues, allows asynchronous and synchronous communications and enables support in varied and appropriate forms of communication, e.g. drawings, text, verbal, etc. to be provided. Finally, the tools and techniques can be readily shared for consistency and efficiency in this environment.

These activities can be carried out in a manual manner, with the team members working in close proximity, preferably in an open-plan office space. However, with the increasing use of computers and their speed and accuracy in many respects of engineering analysis and decision support activities, they can provide essential non-manual support to the team.

(c) Organizational structure

In most organizations, the cultural change to team working will need the highest level of management support to be successful. CE implementation should be taken stage by stage and emphasis should be placed upon changing the company culture.

A sound implementation of CE must be built around the major sequential but highly iterative process, such as the one shown in Figure 1.10.

Matrix management structure has been found to be a suitable form of management organization for the CE approach. Figure 1.11 shows the typical relationships of projects and the major functional areas in an organization.

(d) Quality function deployment (QFD)

This is a formal method primarily designed to accommodate the customer requirements into the design and manufacture of products. The method

CE changing culture

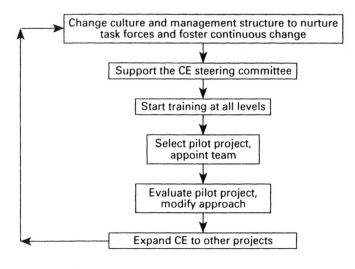

CE should be taken stage by stage with the accent being placed on the changing culture

Fig. 1.10 The CE implementation process.

CE matrix marketing

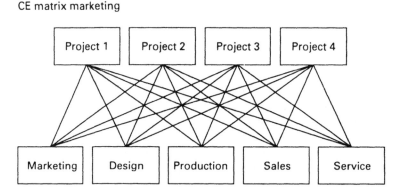

Fig. 1.11 Relationships of projects and traditional functional groupings.

builds in a series of links to the relevant functions in the company for the compromise or resolution of issues. It can also be used to resolve in-house issues in the same manner. As it involves input from a wide variety of sources, it is also a very powerful team and team spirit building tool. This is the topic of discussion in Chapter 5 of this book.

(e) Design for manufacture/Design for assembly (DFM/DFA)

This is also a formal methodology that can assist in building team spirit. The quantitative evaluation approaches used in many methodologies now available necessitates collectively agreeing on scores that need close co-operation of team members from different disciplines. Examples of these systems are the Boothroyd–Dewhurst method and the Lucas DFMA system. Again all the systems are available as software packages. These will be introduced and discussed in Chapters 6 and 7 of this book.

1.4.2 Computer-based support

The computer-based support for CE covers all areas of product design, from market research tools, design tools, manufacturing planning and control tools. The general areas of CE product development of products are:

1 The acquisition and/or development of programs for engineering, design and their management processes.
2 The acquisition and/or development of programs for the communication of information between different computers, programs and locations as well as integration tools.

(a) Engineering design and support programs

These are useful in assisting in the design and manufacture of products. There are a wide variety of programs available of varying capabilities and complexities. They include systems such as computer-aided design and manufacturing (CAD/CAM), and computer-aided engineering (CAE) for electronic, mechanical and other disciplines. Computer-based support initiatives also include engineering data management (EDM) tools, modelling and simulation tools, production planning and costing tools.

The current CAD/CAM systems have the capabilities of three-dimensional shape modelling and the ability to derive physical properties (e.g. weight, centre of gravity, etc.) and to produce manufacturing data such as numerical control data and programme files. Solid modelling provides opportunities to integrate design, process planning, production planning and production through the ability to carry out interactive verification of manufacturing, assembly, and rapid prototyping of parts via newly developed bench-top rapid prototyping machines. These can take a model from a package and produce a plastic prototype in minutes for verification purposes. These facilities enable the whole modelling, draughting and manufacturing process to be speeded up greatly.

Design projects of even small scale can require up to 1000 drawings and there may be up to ten times that of related information. This information is made up of product specifications, simulations, test and analysis

results, costs, schedules, manufacturing, tooling, assembly and maintenance, and so on. The need for EDM tools in CE is unquestionable. A
design and manufacturing audit is the best starting point in the progress
towards integration of EDM tools into a company's computer system. The
audit should determine the following points.

- How the product development process really operates inside the
  organization.
- Which design tools are installed and how well they are currently
  integrated into other systems.
- What systems are currently used by individual functional departments
  and how useful they could be if shared widely across other functions.
- How visible is the use of tools at present in the product development
  process – if not visible, is the product development process a black box
  that supplies answers without a need for real understanding.
- How sequential and repetitive are the various stages of product
  development in the company.
- How difficult is it to obtain and maintain information for products after
  completion of product development.

From this type of information, it can be possible to detail what is desirable
and should be included in the development process in the company. All
information should be registered and catalogued in order to provide
traceability. The company should strive to reduce paperwork, aim for
paperless operation and become once-only data entry in the long term.
Chapters 10 and 12 introduce these systems. They also give practical
examples of commercially available tools and their uses.

1.4.3   Formal techniques

The list of formal methods that are available nowadays is very diverse.
They include systems such as quality function deployment (QFD),
experimental design, Taguchi methods, total quality management
(TQM), design for manufacturability (DFM), design for assembly (DFA),
and design for other 'ilities': productivity improvement techniques (e.g.
JIT, OPT), continuous improvement programs, design maturity assessment tools and costing tools.

   The benefits of using these techniques are well documented. For
example, Digital Inc. have reported that they achieved the following by
using DFM in their transducer manufacturing process:

- total assembly time reduced from 9 minutes 52 seconds to 4 minutes
  and 37 seconds;
- 50% reduction in number of components used;
- 33% fewer assembly operations.

ITT introduced Taguchi methods in 1983 and by 1987 1200 engineers were

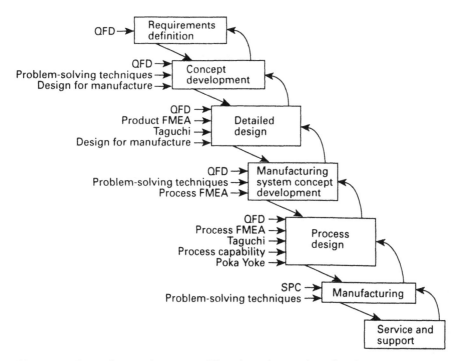

Fig. 1.12 Typical uses of common CE tools in the product development process.

trained in the use of the techniques. A total of 2000 cases were analysed by this method and savings of $35 million a year were made.

The use of formal techniques ensures that there is a consistent approach to the problems throughout the organization. This helps ensure minimum quality levels even with less experienced engineers and designers being involved.

Flexibility is essential in CE, no one technique or methodology offers a universal solution. Choosing the right tools at the right time in the development cycle is an essential skill of the CE team members. Lucas has been an early CE player and Figure 1.12, derived from discussions with Lucas plc (UK) staff, shows how various techniques are applied. Although the computer-based technologies are not indicated in Figure 1.12, it is equally valid for such an approach.

1.4.4 Data interchange methods

As most companies nowadays use computer systems, some extensively, it is paramount that the methods and standards for this activity are available and used. One current method of ensuring that data exchange is possible across the whole company is to ensure that all systems purchased are compatible. However, this tends to tie the company to a

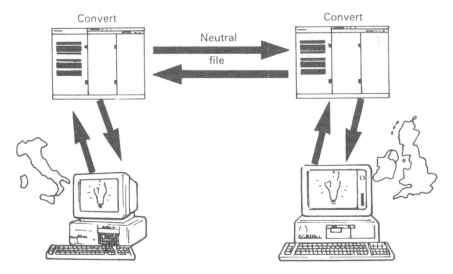

Fig. 1.13 Neutral file format data transfer.

single vendor which can be very risky. Also it is highly unlikely that a single vendor could supply all the computer systems a company may require.

The other option is to ensure that the in-house systems are compatible with the suppliers and customers. This option is only available to large companies who can pressure their suppliers to conform. Otherwise, this approach may mean that the organization and/or its supplier will need different systems for different customers.

The neutral file format route is also available where different systems communicate with each other via a standard format. Figure 1.13 shows how this is achieved in practice. Unfortunately, as the number of systems have increased, so has the number of neutral file formats. The more common standards include:

- IGES (initial graphics exchange specification);
- SET (standard d'exchange et de tranfert);
- VDA (verband der automobilindustrie flachenschnittstelle);
- PDES (product data exchange system);
- STEP (standard for the exchange of products data).

Because of the well-recognized need to focus this diversity into a common international standard, the CALS (computer-aided logistical support) initiative has been started. It aims to provide support for CE via an international standard for product definition data sharing, previously known as STEP. The work towards this standard is still in its draft stage.

A company that is considering implementation of CE would necessarily have to consider a number of factors.

1. The production quantity and volume – high volume or high value products may justify the considerable effort and expense of CE.
2. The complexity of the product – the greater the complexity, the greater the benefits that can be achieved by CE.
3. The expected life span of product – a product with long life expectancy may warrant more efforts in CE than a product with shorter life span. This is due to the possible effect of the technological advances in the area of the products with short life spans.
4. The time to market – if the product can benefit by reduced time to market then it will benefit from CE activity.

1.5 LESSONS FROM SUCCESSFUL IMPLEMENTATIONS OF CE

An investigation carried out by the Institute for Defense Analysis (IDA) USA, identified the following common characteristics in the companies that successfully deployed CE:

- support from senior management;
- changes made are substitutions for previous practices, not additional to them;
- there needs to be a common perception within the organization of the need to change;
- formation of multidisciplinary teams for product development;
- relaxing policies that inhibited design changes and providing greater authority and responsibility to members of design teams.

The key features that can be identified as essential elements for successful CE implementation from the studies are:

- multidisciplinary teams;
- sustained communication and co-ordination across different disciplines and organizations involved with the product;
- use of quality management methods and principles;
- computer simulations of products and processes;
- integration of databases, applications tools and user interfaces;
- a programme of education for employees at all levels;
- employees developed an attitude of ownership towards processes in which they were involved;
- a commitment to continual improvement.

There are four general rules for achieving success with CE. These have been derived from the plethora of published information now written on

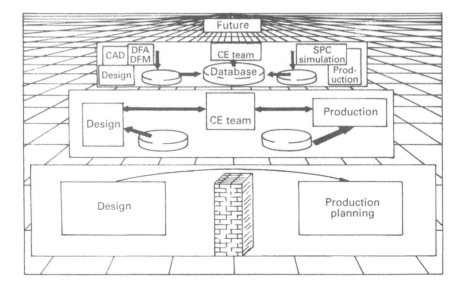

Fig. 1.14   The future of concurrent engineering.

this subject. These rules are only meant to be a guide and need to be set into context in any specific organization.

1. Do not undertake CE until the company is ready.
2. CE deployment is as difficult as the deployment of a major new product line.
3. CE is a methodology and involves cultural change as well as working practices.
4. CE technology, tools and techniques without CE culture will greatly reduce benefits.

1.5.1   The future

The realization of CE requires integration, co-ordination and management of computer-based tools, creation of organizational structure for working in teams, project management, evaluation and the use of appropriate new technologies. The considerations of internal and external customers and suppliers, as well as synchronization of data and information is also essential for optimal results.

Many organizations still operate in a traditional manner. The forward-looking companies are at a stage where CE practice is being tested on individual projects or is becoming normal practice. The activities of the team are supplemented by computer systems and tools. The next generation of frameworks and support tools are being developed by various projects internationally, such as the DARPA Initiative on CE

(DICE) in the USA and a number of European projects funded by EC research and development funding agencies.

CE in the future will utilize tools which will enable team work to be performed in organizations where the various functional expertise is often geographically dispersed. These activities will need to be supported by fast and cheap networking systems, multi-media communications and integration enabling tools, in real time. Figure 1.14 shows the DICE project's view of CE, the 'virtually teaming' capability.

REFERENCES

1. Hartley, J. and Mortimer, J., *Simultaneous Engineering – The Management Guide*, 2nd edn, Industrial Newsletters Ltd., Dunstable, UK.
2. Goldhar *et al.* (1991) Flexibility and Competitive Advantage: Manufacturing Becomes a Service Business. *IJTM* (*Special Issue on Manufacturing Strategy*), 243–59.
3. Watson, G.F. (ed.) (1991) Concurrent Engineering: Competitive Product Development. *IEEE SPECTRUM,* **July,** 22–37.
4. Carter, D. and Baker, B. (1992) *Concurrent Engineering: The Product Development Environment for the 1990's.* Addison-Wesley Publishing Company, Massachussets.
5. Turino, Jon (1992) *Managing Concurrent Engineering· Buying Time to Market.* Van Nostrand Reinhold, New York.
6. *Business Week* (1989), **8** May issue.
7. Ettlie, John and Stoll, Henry (1990) *Managing the Design-Manufacturing Process.* McGraw-Hill, USA.

CHAPTER 2

Concurrent engineering case studies: Lessons from Ford Motor Company Experience

J. V. Chelsom

2.1 BACKGROUND

All the elements of successful concurrent engineering existed more than eighty years ago. They can be seen in the introduction of the Model T, which was developed on a teamwork basis, Henry Ford himself sketching his ideas on a blackboard, and Joseph Galamb and Spider Huff develop-ing them into engineering specifications, while Charles Sorensen checked the manufacturing feasibility [1]. They copied the use of low-weight high-strength steel from the French (a sort of competitor analysis or benchmarking) and they innovated – with the first engine to have all cylinders cast in one block, which was made possible by the idea of separating the cylinder head and the sump from the block, as is still standard practice. Their novel planetary transmission, with its system of pedals and pulleys, was not so long-lived a concept, but it was very close to the principles of modern automatic transmission. Both these ideas progressed through an evaluation process that took account of manufac-turing implications, while seeking ways of surprising and delighting the customer with the product feature. It is not difficult to see the parallels with quality function deployment (see Chapter 5).

Another distinctive aspect of this project, and key to its success, was the way in which customer requirements were predicted through a product vision, a feat of imagination, that anticipated 'the voice of the customer'. This vision was implanted in the minds of the design team, and clearly expressed. Ford's philosophy at that time still seems laudable: 'I will build a car for the great multitude, . . . of the best materials by the best men . . . after the simplest designs that modern engineering can devise . . . so low in price that no man making a good salary will be unable to own one . . . and enjoy with his family the blessings of hours of pleasure in God's great open spaces' [2].

Ford, and most other Western producers of all kinds, lost this team approach, the direct access to the product vision, and the co-operation between designers and producers to meet the forecast needs of their joint end customer. The Japanese never strayed from the co-operative ideal, and succeeded in combining team methods of product development with another American concept – total quality. It was their success that led Ford, and others to rediscover concurrent engineering.

The benefits of CE can be seen in case studies describing recent successes, but the extent of the changes in results and behaviour can only be appreciated by comparison with some of the disasters that emerged from the sequential development processes that were increasingly the standard practice of the Western motor industry from the 1920s to the 1980s.

The first case studies given in Sections 2.2.1 to 2.2.6 are therefore examples of these bad old ways – aptly called 'over the wall' engineering, where the designer works in isolation until something is ready to be tossed over the wall to the process engineers who either toss it back or develop a process which is tossed over to production.

The need for better quality and shorter product development time spans was generally recognized, but actually achieving these objectives turned out to be difficult. This is illustrated by two of Ford of Europe's projects from the early 1980s described below – one relates to a new engine, the other to a transmission.

2.2 TWO OVER THE WALL CASE STUDIES

Strategic review – the expansive name for strengths, weaknesses, threats and opportunities (SWOT) analysis – had shown that Ford product quality generally was not competitive in Europe, and certainly not 'best in class', which was the corporate objective. It also showed that the most serious product weakness was the power train – engines and transmissions.

In 1983, decisions were taken to rectify this situation by replacing the mid-size range of engines, which were single overhead camshaft eight valve products made in both Cologne and Dagenham. Relative to competition, the engine had poor power and economy and was noisy and unreliable. The mating four-speed transmissions had poor 'shiftability' and were heavy. The replacements were to be a double overhead cam engine and a new five-speed gearbox.

2.2.1 Programme assumptions

The assumptions for the engine programme were:

- volumes – 240 k per annum;
- design – in line four cylinder DOHC, eight, 12 or 16 valves;
- material – cast iron block and head, pressed steel sump.

Manufacturing cost comparisons showed that the new engine should be built only in Cologne, with a job 1 date of July 1986.

For the transmission, which would be matched with other engines, besides the new DOHC, the assumptions were:

- volumes – 640 k per annum;
- design – one piece housing, six variants for six engines;
- material – aluminium.

Repeated checks with product planners in Truck Operations confirmed that the new transmissions would not be fitted to the new Transit van. Production of the transmission case and clutch housing was to be in Halewood in the UK, with assembly in the UK and Germany and with job 1 completion in May 1987. A simplified timing plan for the two programmes is shown in Figure 2.1.

2.2.2 Engine programme implementation

A critical element affecting the programme was that in September 1983, when enquiries were made for the long lead equipment for the engine, the world machine tool market was at a depressed level. Also, the most successful suppliers to Ford of this equipment (transfer lines to machine the cylinder block and cylinder head) were US firms anxious to establish a presence in Germany. These two factors combined led to very competitive quotes from these two firms – good technology, and prices below the previous Ford engine programme in 1977. Release of funds in March 1984 meant that orders could be placed for prototype parts (using a new form of commitment for castings – a result of studies of 'the way the Japanese do it') and the block and head lines ordered with delivery to be completed in September 1985. In May 1984 the orders were put on hold, to allow review of the 'cycle plan' – the forward vehicle model programme – which could affect the model application, and hence the volume requirements for the engines. The job 1 date was revised to July 1987. The machine tool suppliers were asked how long the restart instruction could be delayed without affecting price or the new delivery timing. The answer was 'November 1984'. The 12 valve version was deleted in July 1984. Funds were unfrozen in November 1984, but the Ford designers had been busy in the meantime, and released new part drawings showing the engine to be longer and higher. Possible additional vehicle applications meant that

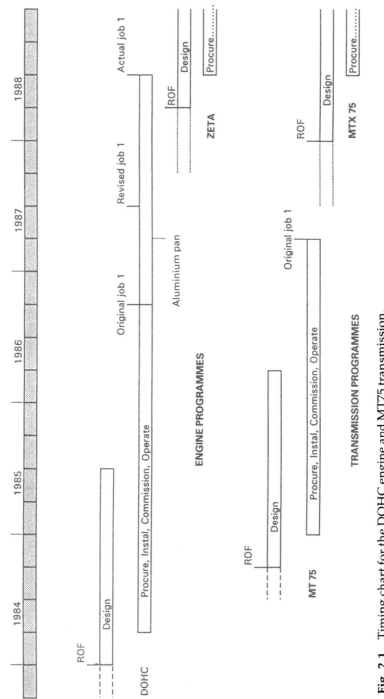

Fig. 2.1 Timing chart for the DOHC engine and MT75 transmission.

provision had to be made for expanding output to 450 k a year. Evaluation of these changes resulted in the suppliers being unable to proceed until January 1985. In March 1985, it was decided that the new engines should be built in Dagenham, not Cologne – improvements in UK efficiencies and exchange rate movements made this a lower cost location. Delivery timing requirements were unchanged, though job 1 moved a month to August 1987 because of summer shutdown timing. There were many major design and process changes, but the last machine tools were delivered on time by November 1986.

2.2.3   Transmission programme implementation

Meanwhile, the major equipment for the transmission had been ordered – 'design only' in December 1984. Requirements were very demanding: to meet performance objectives, the case had to be machined with some critical holes on the perimeter of a bell-shaped housing located with tolerances of ±1.5 microns between centres. (See Figure 2.2.) To achieve this, a pallet transfer machine was proposed, with every one of 166 pallets identical, and made with location points varying by not more than 0.01 mm. Operating temperatures during metal cutting had to be controlled to 20 degrees C ±1 degree. Production rates were 200 pieces an hour. To provide this volume and the flexibility between the six types, a tool change time cut-to-cut of 20 seconds was required, and a complete changeover of 3 minutes. Separately, these are demanding requirements. Combined they represented a new order of precision, speed and control.

Release of funds allowing complete commitment of the order was made in March 1985. At this point, although no drawings were available, it was made known that an additional version of the case would be required so that the transmission **could** be fitted to the new Transit, which reversed a frequently checked programme assumption. At the end of April drawings were supplied, but they were all different from those that had been used for quotation and subsequent preparatory work. The three transfer machines became four to handle the new volumes and variety. In May 1985 it was announced that the case would be made in Germany – the reverse of another key assumption.

2.2.4   Problems with the engine

The engine programme, with a great deal of effort by Ford and supplier personnel, was proceeding on time. In February 1987, just a few months before start of volume production, a note was issued describing 'Application of CIM to the DOHC engine'. It was all there – as shown in the plant layout – electronic data transfer from design engineers in Cologne to process engineers in Dagenham, JIT, automated material handling, SPC, simulation of operations on computer etc. (see Figure 2.3). But in March it

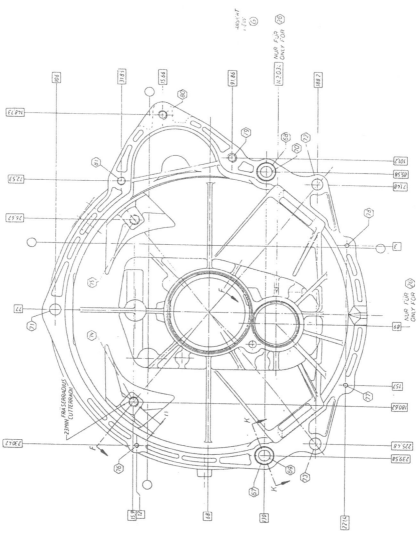

Fig. 2.2 Critical dimensions on the MT75 transmission case Centres of holes 67 and 68 located within 1 micron

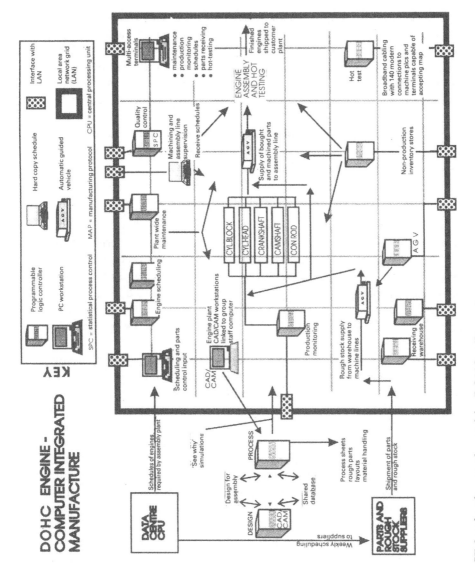

Fig. 2.3 CIM technologies planned for DOHC at Dagenham

was decided that the engines would not be fit for sale. Tests of engines using parts from the volume production facilities and operators showed a number of concerns, notably oil leaks from around the sump/block interface. This was because the steel stampings from the company's own press shop, made at the rate of hundreds per day, were not as consistent dimensionally as those used in prototype testing, which came in low volume from outside specialists. The solution was a cast aluminium sump, which of course needed new foundry tooling, new cutting machines and major changes to the assembly equipment. This all took time – in fact a whole year. It also cost a lot of money.

2.2.5   Problems with the transmission

The transmission programme was never thought to be proceeding smoothly. In February 1986, to the machine builders' surprise, tryout parts were delivered with added ribs. It was explained that these were no concern for processing – the ribs were to counter noise and vibration evident from prototype tests. More ribs were added, and a 12 kg damper fitted to the drive shaft before these concerns were overcome. The 12 kg was more than the weight saved from the redesign of the transmission case.

The ribs were one more complication for the machine tool company, who were finding many interesting facts from their trials. Parts that were OK in the machining fixture were not OK in the inspection gauge. Parts that were OK in the inspection gauge were not OK half an hour later. The Ford process engineers eventually recognized that the cutting machine was more accurate than the gauge, and that temperature changes in normal factory conditions generated more dimensional variations than material and process variations. It was not until February 1987 that a design engineer visited the machine tool builder to discuss these concerns. By then the subject was so sensitive that he and almost every other engineer retreated to stand off positions, insisting that original specifications should be met, and each party was to solve its own problems. Personal relationships between top executives at the supplier and customer ends became so strained that they too were a cause of delay. Eventually there were nine variants of the transmission case, and more than 220 changes of design or process affecting cost and timing.

All these concerns with the case masked other difficulties with other components and processes, such as laser welding of the cluster gears, and electro-chemical machining of a synchronizer. These also suffered from the stand off between design and process engineers – the same people were involved once 'discussion' went to management level.

2.2.6   Market and cost impact

The engines went into production in August 1988. For almost two years a huge investment had stood idle, and carefully trained operators had drifted away to more rewarding jobs. Equipment costs had been pushed up by all the changes, and for the three major items alone this amounted to more than $40m. The suppliers all lost heavily – in the region of $5m for each of the two transfer line vendors. Component suppliers suffered too – many of them had made investments in design, facilities and training to supply new parts, which were not required for a year or more.

The outcome on the transmission programme was similar. Ford suffered idle plant, lost learning, escalating facility costs and late start of production. It was the same for equipment and component suppliers.

It was not until early 1989 that the DOHC engine and the MT75 transmission were offered for sale together in a vehicle. By then the competition had moved on, and far from being 'best in class' state-of-the-art products they were just another power train.

The lessons from these case studies are:

- Planning assumptions should be clear and consistent.
- Solutions to design and manufacturing concerns, especially those involving new materials or process innovations, have to be pre-tested using parts representative of production components, and the results shared by designers and producers.
- Goals have to be shared and inconsistencies removed.
- Quality should not be compromised by short term cost considerations.
- Communications have to be open and free from fear, using well developed interpersonal skills.
- Product planning, design, process engineering, manufacturing and suppliers are interactive functions – barriers between them have to be removed.
- Complex equipment requires targeted training, quickly put into practice.

2.3 CONCURRENT ENGINEERING CASE STUDIES

The inadequacies of 'over the wall' engineering for innovation and the management of new technology have been exposed. It would be reassuring to see that the lessons have been learned, but all around, even in the 1990s, there is evidence that 'the voice of the customer' is not being heard, and that products are being tried and tested by the customer rather than the supplier – or better, by the supplier and the potential customer. The need for improved ways of managing new product programmes was recognized by many Western companies early in the 1980s, because of their poor performance in world markets in the 1970s. Concurrent engineering was the result of this search for a better way, though it can now be seen as a re-invention. Despite its clear advantages, CE has been slow to become re-established, because of the difficulties of changing entrenched attitudes and dismembering bureaucratic organizations.

General Motors was one of the first companies to react in 1982, with re-organization of their product engineering and manufacturing engineering activities in the USA, bringing them together to work in two huge teams – one concerned with current or imminent products, the other to work on future products. Resistance to changes on such a scale was so great that they were never fully implemented, and have been overtaken by other proposed solutions to General Motors' even greater problems in the 1990s.

Ford's search for better ways of doing business was prompted by the dramatically shorter lead times and lower capital costs that in 1979 enabled Mazda, the Toyo Kogyo company in which Ford has a 25% stake, to win a contract to supply transaxles to Ford US, after bidding against Ford's European operations and outside suppliers in the USA. Executives from Ford of Europe were sent to Japan to find out how Mazda could be so quick at such low cost. Although lessons were learned across the whole spectrum of car design and production, the ability of Mazda to produce representative prototype engines and transmissions in about six months really captured Ford's interest. It took them that long to buy prototype castings! Pressure from Ford's engine designers to match Mazda in this area led to some of the earliest examples of successful SE by Western automotive producers.

2.3.1 CE Case Study No. 1: The 'ZETA' engine programme – a new way of doing business

In 1983–4 Ford's European operations began 'pre-sourcing' castings for new designs of cylinder heads and blocks. This involved identification of the future full volume production source on the basis of past performance, and committing to buy prototypes from them when the design was 'just three lines on a piece of paper'. With minimum information about materials and dimensions, the selected foundries were able to schedule pattern-making and production without queuing time, and without going through the enquiry/quote/select process of competitive bids. Prototype lead times were halved overnight!

This small-scale success contrasted starkly with delays and cost escalation due to late changes in the DOHC engine and MT75 transmission programmes, and led to the idea that similar changes of procurement practice could be applied to selected high-cost, long-lead components and equipment. By late 1985, the new approach was being discussed with potential suppliers for the 'ZETA' engine – a four-cylinder double overhead camshaft engine to be developed in Europe for manufacture in North America and the UK. At that time it was planned that only a new aluminium cylinder head would be designed, to be fitted

ZETA ENGINE
OBJECTIVES

1. Earlier prototypes

2. Prototypes more representative of production units

3. Prototypes built by production suppliers/plant

4. Engineering releases more representative of final product

5. Earlier resolution of feasibility issues

6. Elimination of enquiry/recap phase

7. Joint prototype/validation programmes

8. Equipment cost reduction

9. Piece cost reduction

10. Quality improvement

Fig. 2.4　Objectives set for the ZETA task force.

to existing cast iron cylinder blocks used in the different US and European 'Erika' engines – which had also started as a common 'world' design, to be made by common processes, but which ended up with only one component similar enough to be made on almost identical equipment. For ZETA, the European design office was designated the 'centre of responsibility' – to design in one place a product to be made and sold globally. Similarly, the European Engine Manufacturing Engineering department was assigned responsibility for development of production processes, and the purchasing activity took on the task of sourcing components and equipment.

(a) Supplier involvement

Supplier selection reflected this requirement for capability to support production on a global basis, and the group brought together for the inaugural meeting in April 1986 comprised Montupet Foundry (France) for the casting, Lamb Technicon (USA) for machining, and Comau (Italy) for cylinder head assembly. Each supplier was asked to bring their one task force member to meet the one member from each Ford function. The brief for the task force is shown in Figure 2.4. The first meeting was a shambles, with half the Ford design engineers saying it was a year too soon, and half that it was a year too late, but it did succeed in identifying the task force leader, a young German design engineer, Dr Rudi Menne, and setting a date for the first working session.

Ford of Europe Incorporated

J. V. Chelsom
Director – Facilities and
General Supplies – Supply

11 July 1986

Mr P. Cantarella
COMAU S.p.A
Via Rivalta 30
10095 Grugliasco
Torino

Dear Paolo,

All the Ford activities working on the 1992 Zeta engine programme are very pleased that Comau are joining us in a new partnership approach to development of the product processes and equipment involved.

The aim is to resolve design and manufacturing concerns before equipment is ordered, and so avoid changes later in the programme with their inherent threats to quality, cost and timing.

Based on our past co-operation, and your expertise in this particular field, we believe that Comau will make a major contribution to the success of this innovative business relationship. Recognizing this, it is Ford's intention to order from Comau, without competition, the automated assembly equipment for the cylinder head, both within the transfer line and following on from it. We both recognize that there are risks as well as benefits in this enterprise.

There is the possibility of initial disagreement on price but, as in the past, this could probably be resolved by negotiation. Only if this failed would we seek competitive bids, possibly using ideas contributed by Comau. It is also possible that the programme will not be approved, or that after approval it could be postponed or cancelled.

Should these or other events prevent us from completing the expected orders with you, Ford would discuss ways of compensating Comau for the costs incurred through your special involvement over and above your 'normal' sales and proposal expenses.

Yours sincerely,

John Chelsom

```
COMAU


                              Grugliasco, Septembre 15, 1986
                              034/rm



Thank you for your letter on the 1992 Zeta Engine Programme.

    We are very proud of having been chosen by FORD for the
assembly equipments.

    In principle I agree with your proposal as described. I am totally
confident of FORD and you personally, so in any event we would
find together solutions able to protect Ford and Comau's
interests and right.



                                            Paolo Cantarella
```

Fig. 2.5   Letter exchanges between Ford and equipment suppliers.

At a later stage, the group was expanded to include Krause (Germany) for total engine assembly and Ingersoll Milling (USA) for block machining. The brief was to jointly design an engine that would be considered 'best in class' by customers worldwide with regard to:

• DQR – durability, quality and reliability;
• NVH – noise, vibration and harshness;
• P and E – performance and economy;
• performance feel – you know it when you feel it!

The engine was to be made at minimum production cost, with minimum investment.

Volume production was due to start in the UK in mid 1991, based on programme approval and release of funds in mid 1988. Detailed involvement of suppliers so far ahead of financial authorization was unprecedented, and broke many Ford house rules. It required a completely new form of agreement, which was developed between the Ford purchasing director and the heads of the equipment supplier companies. This was put into effect by a simple exchange of letters, as shown in Figure 2.5, but the spirit of co-operation was so strong that the letters were signed after work had started.

ZETA ENGINE TASK FORCE GROUP

Objectives

1. Japanese approach to achieve best in class new medium I4 engines.

2. Corporate engineering (Product Development Manufacturing, Supply). Earliest involvement of all areas.

3. Optimized design to achieve customer's satisfaction at minimized production costs.

4. Improved communication to recognize concerns and establish solutions in the very early phase of the program.

5. Ownership of the product within all areas (throughout the total Corporation).

Fig. 2.6   The task force's own version of their objectives.

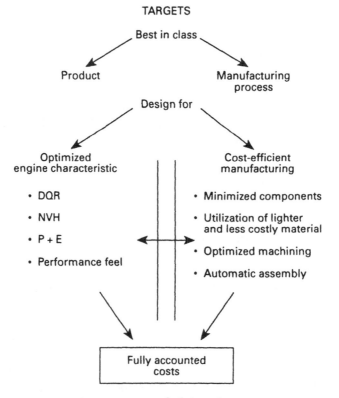

Fig. 2.7   The task force's own route to their targets.

DESIGN OPTIMIZATION STRATEGY

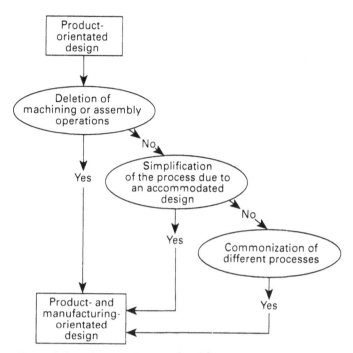

Fig. 2.8 The task force's optimization algorithm.

(b) Getting started

The task force met initially in Cologne for three days, to decide their own operating rules and work plan (Figures 2.6 and 2.7). Their self-education quickly took them to Turin to see a complete transfer line on the Comau shop floor. This was a 'first' for some of the designers. Then they went to Dagenham to see the final stages of DOHC engine equipment installation – another revelation! This not only led to better understanding of the ZETA task, but the shared experiences began to change the group from a task force to a team.

The study processes were recorded simply, by notes and sketches such as in Figure 2.8.

A list of opportunities was developed (Figure 2.9) and evaluation was assigned to sub-groups, who again used simple documentation such as in Figure 2.10 which relates to a study led by Lamb.

Overall, the team followed a classical approach:

- identify, agree and quantify objectives for customer satisfaction;
- cost reduction through elimination of operations, simplification of operations, or commonization;

CYL. HEAD DESIGN
Items under discussion

1. Oil drain holes to be cast finished
2. Elimination of Tappet Lube Holes
3. Cam-bearing lubrication
4. Deletion of machining for foundry core holes
5. Bolt holes to be cored finished
6. Cam cap mounting holes
7. Deletion of positive valve stem seal spigot machining
8. Injector bores in the manifold, not the cyl. head
9. 2V and EFI castings to be common
10. Common valve sizes between derivatives
11. Assembly of finished machined seat and guide inserts
12. All lobes on one camshaft to be common

Fig. 2.9   List of opportunities.

CASTING OF OIL GALLERY AND LUBE HOLES
Sub-study by Lamb

Feature

Longitudinal galleries cast in, complete with cast tappet feed slots also serving as core supports as required

Advantage

Eliminates all machining for tappet feeds

Disadvantages

Risk of core breakage for high volume production

DQR risk due to difficult cleaning of cast galleries

Increased oil level in cyl. head (aeration, pumping losses) requires restrictor valve

Increase start-up noise even with check (non-return) valve

Fig. 2.10   Sub-study outline.

- brainstorming to establish a list of opportunities;
- delegation of opportunity evaluation to expert sub-groups;
- evaluation of preferred alternatives by build and test;
- use of directional cost estimates to make stop/go decisions.

(c) Simple controls and hands-off management

Another example of the simple but effective documentation is shown in Table 2.1. It is worth noting that although state-of-the-art data processing facilities were available, and were used for actual design and process development, communication between the team members was by hand-drawn sketches and typed notes. The core team got together as required – usually once a month for a few days – and in between got on with their sub-studies (and work on other programmes) in separate locations which were hundreds of miles apart. The group's face-to-face contact and team spirit was far more effective than computerized data exchange between strangers – which is a risk now with the more advanced communications capability available.

The team was not required to make special management reports or presentations. Copies of their routine review meeting minutes were sent to the three sponsoring directors, and there was an open invitation to call on these directors for decisions or resources when the team thought necessary. The directors were brought together only once – to hear results!

(d) Quality, timing and cost benefits

By the time of programme approval in June 1988, the interaction of team members, rather than the efforts of any individual, had led to potential investment cost reductions of more than $10m. This was built into the purchase order on Lamb for the cylinder head transfer line. Continuing co-operation throughout the programme led to even more significant savings through the avoidance of change. Though 'non-incurred' costs are impossible to measure, some idea of the benefits can be inferred from the immediately preceding engine programme – the DOHC project described earlier, in which costs for the three major pieces of equipment escalated by $40m, or more than 60% (see Table 2.2). It has to be stressed that even with these additional payments the suppliers lost heavily on their contracts. This was demonstrated by open access to their audited accounts and internal cost information. This is quite contrary to the nonsense about vendors making their profits on changes with buyers' connivance, as suggested in *The Machine That Changed the World* [3] – which according to the female half of the population should have been the washing machine, not the car.

Table 2.1 1992 Zeta engine programme design for competition (cylinder head)

#	Design feature	Advantage	Disadvantage	Piece cost effect	Investment effect (5000) max	Action responsibility timing	Concurred as prime programme
1 i	Oil drain holes cast finished	• Machining operations reduced • Tool cost reduction	• NAAO comment – increased fingers on oil jacket core which may result in increased core breakage • Montupet state no change in cost		(400) × 2	• Oil drain holes on exhaust cast or machined may result in water circulation problems around exh port • PDG to review	YES
ii	Self-contained hydraulic tappets	• Machining operations reduced • Tool cost reduction • Minute cost reduction • Product improvements claimed by tappet supplier	• Tappet design not fully developed (DQR concern) • Piece cost increase • Longer assembly leakdown required • May require additional squirt holes for lubrication between tappet and valve stem • No fallback route in the case of failure if oil hole machining is not protected	T B E	(4000)* × 2	• Action plan established • Resource and facilities to be established • EAO/NAAO joint testing	TBA
iii	Delete camshaft lubrication holes in half bore	• Machining operations reduced • Tool cost reductions	• External system for Cam bearing lube required DQR • Piece cost increase	T B E	(600) × 2	• Design study to be established • Build prototype for test	TBA
iv	Delete machined oil gallery system by casting the oil gallery and tappet lube holes	• Machining and assembly ops deleted • Tool cost, minute cost and floorspace reduction • Reduced cast weight	• Increased foundry costs • Increased oil in the cyl head may require check valve	T B E	(4000)* × 2	• Foundry feasibility under review • Montupet/NAAO • Design study required • Build prototypes for test	TBA
2	Delete machining of foundry core holes Seal with gaskets, manifolds etc (reduced 8 holes to 1 only)	• Machining and assembly ops reduced with the following benefits • Piece cost reduction • Tool cost reduction • Minute cost reduction • Reduced production scrap	• Foundry concern with regard sand removal, degas and core support • Wall thickness variation from 0 5 to 0 8 • DQR concern (plus and minor)	T B E	(1200) × 2	Series I heads will be made to new level	Montupet YES NAAO TBA
3	Core bolt hole – use dowel holes for head/block location	• Machining operations reduced • Tool cost reduction • Minute cost reduction	• Design concessions required for foundry draft angle, diameter tolerance, positional tolerance • Reduced sealing land – prevent design requirement of not connecting bolt boss to the cyl bore walls thus eliminating cyl bore distortion		(650) × 2		NO

Table 2.2   Cost escalation on the DOHC engine, with OTW design

	Number of major changes	Escalation in cost (%)
Cylinder head transfer line	37	59
Cylinder block transfer line	19	68
Engine assembly equipment	19	123

Although these three major lines and all other machines for the DOHC were available on time, the engine was two years late reaching the market, which was quite unexcited by its arrival – another immeasurable loss through missed sales opportunities and inability to price for a new premium product. A conservative cost estimate of keeping the whole investment inactive for two years would be $40m, so with the $40m cost increases and lost sales, 'over the wall' engineering probably reduced profits by at least $100m.

By contrast, changes and cost increases on the ZETA engine were less than 5%, and the engine was available in line with the original timetable. This timetable allowed validation and demonstration of capability for all operations prior to the start of volume production, so that world class quality was achieved from job 1. In fact, all quality, cost and performance objectives were met or beaten. And the equipment suppliers made profits!

2.3.2   CE Case Study No. 2: The MTX75 transaxle

(a) Spreading the word

It was intended to use the ZETA programme as a pilot for the new practices that came to be known as concurrent engineering, and to compare results with a Ford US modular engine programme. This started later than ZETA, and gathered ideas from Europe, to which others were added with the aim of tighter timing and cost control. However, the early signs of success in Europe were so encouraging that the concepts were applied more widely, long before the two engine programmes had run their course and a single 'best practice' had been developed.

From ZETA, the first extension of CE in Ford Europe was to the MTX75 transaxle programme – again under pressure resulting from adverse experience on the conventionally managed MT75 programme described in Section 2.2. The MTX75 is an adaptation for front wheel drive of the MT75 transmission used on rear wheel drive vehicles, so the only unavoidable change involved the manufacturing of the transmission case and clutch case. However, concurrent engineering was used to make other improvements. There were three related studies: one to develop the essential changes to the housings; another to redesign some elements of the gear train to facilitate completely automatic assembly; and the third to introduce new gear finishing processes.

(b) Progressive development

The assembly process was developed first in a test cell, and all three projects were implemented in three further phases, namely the pilot programme (low volume), the main programme and the capacity expansion. This progressive approach enabled new materials and processes to be fully evaluated and concerns to be eliminated well ahead of job 1 production dates. Without CE these quality and cost advances could not have been achieved within the timing required by the corporate model cycle plan.

The pilot programme was needed quickly so that the transaxle could be fitted to a special 'sporty' car that had an important marketing role. The main programme was for a new high volume car, and the expansion for a new larger car two years later. The objectives for the three projects were the same:

- customer satisfaction;
- best in class quality;
- reduced design-to-market launch timing;
- minimum design and process change in mid-programme (no slack for this in timing plans);
- reduced investment cost;
- high equipment carryover between phases.

Team members were drawn from somewhat different activities for the three studies, as shown in Table 2.3. The involvement of the foundry and diemaker from the outset, when prototypes were made from sand castings, was a good example of advanced quality planning. They were able to advise on design and to plan their processes for volume

Table 2.3   Activities represented in the three MTX75 CE teams

Activity	Case machining	Assembly	Gear finishing
Design engineer	x	x	x
Process engineer	x	x	x
Equipment suppliers	x	x	x
Gauge/Test suppliers	x	x	
Foundry	x		
Diemaker	x		
Logistics consultant		x	
Systems office		x	

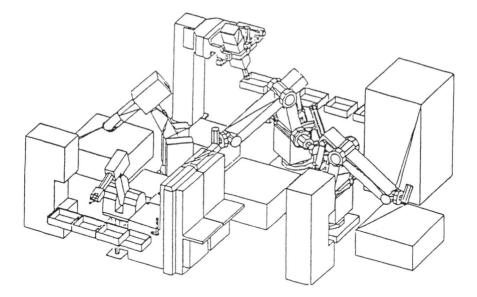

Fig. 2.11 Robocad cell.

production of high quality pressure die castings, so avoiding delays through doing this work sequentially later in the programme, with a risk of redesign then being required.

Use of Aachen University as logistics consultants in the automated assembly project enabled material flows, inventories, plant layout and process capacities to be optimized before capital investments were made.

The Systems office not only provided the usual CIM support, they also introduced ROBOCAD which generates on-screen simulation of robot paths (Figure 2.11). This off-line evaluation minimizes cycle times and optimizes cell layouts much faster than physical experiment, and prepares control system software at an early stage – another time saver compared with add-on programming late in the project (Figure 2.11).

(c) CE for machining operations

The team working on the transaxle housings developed designs and processes suitable for production on machining centres for the low volume phase, and on transfer machines for high volumes. Significant investment cost savings were made, but the greatest benefit was that the low volume facility was available one year earlier than would have been possible with conventional project management and transfer line production. This meant that the vehicle with the new transaxle could be launched a year earlier to take its place in the marketing plan.

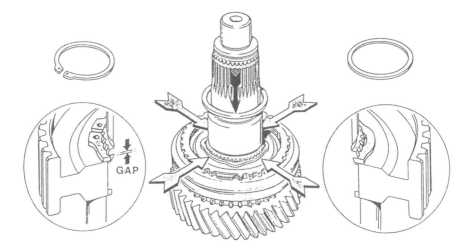

Fig. 2.12   Zero-play retaining ring.

(d) CE and design for assembly

The automatic assembly team first concerned themselves with redesign for assembly rather than simply automating assembly of the existing components. This reduced the number of parts from 70 to 49, and the assembly cycle time by 30%. A sub-study of a retaining ring provided an outstanding example of the benefits of CE:

- FMEA showed a small, unacceptable risk using a split ring.
- **Together**, the design engineers, process engineers, part and equipment suppliers developed a ring whose design, dimensions and material allowed deformation in the assembly process to provide an absolutely secure, zero-play fit (Figure 2.12).

Again, the team approach of CE brought the product to market earlier, with higher quality and reliability, and lower piece cost. Investment costs were reduced by $9m at the high volume phase, as a result of optimizing processes through experience of the lower volume phases.

(e) CE and quality planning

The third MTX75 CE team was brought together as a result of early quality planning and process FMEA. The design requirement for additional torque necessitated a compact set of hard-finished gears for the final drive. Difficulties were foreseen due to:

- gear distortion after heat treatment;
- limited stock on pinion bandage;
- risk to finish in press fitting the pinion;

- no possibility of gear finishing after fitting the pinion.

Various processes were considered – grinding, skiving, honing – but none guaranteed the quality required. The Hurth company of Munich was invited to join the team after a preliminary meeting to explore their willingness to work on the same basis as had been developed for the ZETA equipment suppliers. (There was no possibility of the normal type of enquiry or order, since there was no known process or machine.) Hurth's specialist knowledge of gear manufacture and gear finishing machines, combined with flexibility and an open-minded attitude from the Ford engineers, enabled a solution to be found. The quality capability of the new process was validated within the tight timing required by the success of the other two MTX75 CE teams.

2.4 LESSONS FROM THE CE CASE STUDIES

There are some common factors in the ZETA and MTX75 examples that are key to the success of any CE project.

- Early, careful selection of team members.
- Component and equipment suppliers must be selected and involved early enough to participate in the design process.
- The team must focus on the quality of the end product, and achieve quality targets from the start of production by validating every component and every process during development.
- Volume producers must be involved in prototype production to identify as early as possible any opportunity to improve process reliability.
- New processes must be validated at the prototype stage, if necessary using surrogate production-like parts.
- All parties must be able to share information from prototype work – this must not be the designers' secret.
- Personal attitudes are more important than computer hardware for effective communication, and training in group problem solving and helpful team behaviour may be needed.

It is difficult enough for the CE teams to learn these lessons, but even more difficult to spread the learning throughout the whole company. In many companies there are long-established organizational and physical barriers between designers of the product and designers and users of production processes. This leads to mental barriers which are an even greater obstacle. These internal barriers also contribute to the difficulties of learning from customers and suppliers outside the company – if only sales people are allowed to talk to potential customers, the designers will never hear what the customers really say. Similarly, if only purchasing people are allowed to talk with and listen to suppliers, the product and process designers will be cut off from this enormous pool of expertise.

The removal of mental barriers through training, and above all by leadership example is the most serious challenge to the successful introduction of simultaneous engineering. The best prescription for success comes from Roy Wheeler of Hewlett Packard:

> What tools does an engineer need to get started in CE? Pencil, paper, some intelligence and a willingness to work with peers in other functional areas to get the job done [4].

REFERENCES

1. Lacey, R. and Little, T. (1986) *Ford – The Men & the Machine*, Brown & Company.
2. Sussman, I., Warren, and Lacey, R. (1985) *op cit*, quoting *Culture as History*, Pantheon Books, NY.
3. James, P.W., David, T.J., and Daniel, R. (1990) *The machine that changed the world*. Rawson Associates, NY: Collier Macmillan Canada, Inc. Toronto.
4. Watson, G.F. (1991) Concurrent Engineering, Special Feature of *IEEE Spectrum*, **July**.

CHAPTER 3

Organizational and managerial issues

K. S. Pawar

3.1 INTRODUCTION

In many industries, until a few years ago, the design of a product and its subsequent economic manufacture was typically the responsibility of one individual; today this is true only in smaller companies. Here the designer is expected to have a sufficiently broad background so that a design can be produced which is sound functionally and can be manufactured economically in the desired quantities. Thus the designer performs a dual role of product designer and production specialist.

The product design process requires a team effort both to decide on what product to develop and how to provide effective solutions. The team leader has to co-ordinate all the specialist support and use his or her personal managerial and technical judgement to decide between the alternatives with which the designers are confronted.

The main objective of a design team is usually to produce products that offer maximum value to the customer at the least cost to the company. The team needs to be aware of the performance features and combine these with cost constraints imposed by the resources available for the particular project. The product specification therefore needs to make a major contribution to the cost-effectiveness of the design project by reflecting the realistic needs of the market.

Once customer needs are satisfactorily defined, the design problem falls directly on the development team. Here multidisciplinary specialisms, such as electrical or mechanical engineering, are often needed. Since several specialists may be involved during the design process of a product, it is imperative that their efforts are all harmonized effectively if the company is to be competitive.

Due to the existence of a group of specialists in the organization many different interfaces may become important. These include the interface relationships between management and design in determining the

technical environment, the commercial/design interface in the specification of need, the design/development interface, the influence of
standardization policy on design procedure across the design/
manufacturing interface and so on. In fact the list of interfaces can be very
long indeed and will be determined by the size of the company and
product complexity [1].

The problems which these interfaces create tend to be obscure and are
sometimes difficult to define; solutions to them cannot always be found in
quantitative evidence. The ultimate conclusions therefore have to be
made from the qualitative information available from industry.

3.2　ACHIEVING INTEGRATION

Gillen [2] believes that the expansion convergence process is crucial for
the implementation of CE. Organizations usually have an abundance of
specialized knowledge but lack convergent knowledge. Convergent
knowledge gives separate functions the ability to synthesize their
expertise to accomplish interdependent tasks. Each year, organizations
spend large amounts of time and resources attempting to converge the
specialized talents and knowledge in their firms to produce quality
products on a timely basis.

Expansion convergence is the process of combining diverse organizational talents and knowledge into collaborative efforts that solve
problems and take advantage of opportunities. Convergence involves
taking specialized knowledge from multiple sources and forming a new
collective mindset that allows people to work together at a much deeper
level. Despite the recognition of the importance of integration, high-
technology firms frequently lack a systematic approach to this problem.

Convergence can only occur when each functional group understands
the needs and goals of the others as well as of the business as a whole. The
Digital Equipment Corporation, for example, shortened its product
development time, reduced product costs, met time-to-market requirements, and increased quality by encouraging co-operation across engineering disciplines.

Expansion convergence takes Digital's efforts one step further by
applying the concept of concurrency to the whole organization, not just to
the engineering process. In order to achieve concurrency, a firm has to
develop convergent knowledge, which requires bringing all of an
organization's specialized knowledge together in creative new forms.

The expansion of knowledge allows individual departments to converge their efforts with the rest of the organization. An organization's
parts must expand before the total organization can converge. For
example, a personnel department that understands the essence of the
core business can transcend the basics of its function and contribute

Table 3.1 Methods of promoting design-manufacturing integration

Cases	Design-manufacturing team	Compatible CAD system for design and tooling	Common reporting position for computerization	Philosophical shift to design for manufacturing	Engineering generalists
Off-road vehicles	x	x	x		
GE steam turbine			x		
Auto division	x			x	x
Xerox copiers	x			x	
Tools	x				
Appliance division				x	
Rockwell Space		x	x		
Allen-Bradley	x		x		
Amana	x			x	

through such programs as organizational development, human resource strategies and management development. Engineering departments that expand their knowledge can begin to grasp the financial costs of their activities and provide key insights into how to manufacture better the products they design.

Manufacturing can work closely with engineering and marketing to ensure that designs are capable of being manufactured and to make sure they incorporate customer requirements. Finance can become internally directed through the development of systems and decision modelling, which enable the organization to measure the costs of quality and trace the profitability of strategic decisions. The accounting department can be expanded beyond its typical functions. Accounting systems can be enhanced to include data that show the real costs incurred in developing and producing a product [3].

The essence of concurrent engineering is concurrency and integration. There are various methods for achieving integration. Ettlie and Stoll [4] highlight emerging patterns to integrate design and manufacturing as part of the effort towards CE at various companies. Table 3.1 illustrates this point. It can be seen from Ettlie's work that different companies used different combinations of methods to promote the integration of design and manufacturing, that is, those combinations that are suited to their situation and needs.

Trygg [5] has managed to show that companies which have been successful in shortening their time-to-market also put different emphasis in the use of multifunctional teams, computer integration and analytical methods and tools (see Table 3.2). Similarly, the work of Hunt [6], summarized in Table 3.3, also shows the different levels of emphasis made on multidisciplinary teams, software solutions, design production processes, computer systems integration and so on. Thus any strategy to implement CE must choose the right mix of tools and methods and the appropriate stages for their application. Even though different companies

Table 3.2 The emphasis on the generic elements as key factors

Company	Multifunctional teams	Computer integration	Analytical methods and tools
ABB	Ss	S	–
AT&T	S	S	S
British Aerospace	Ss	m	–
Digital Equipment	Ss	S	S
General Motors	Ss	–	–
Goldstar	S	S	–
Honeywell	Ss	–	–
Hewlett-Packard	Ss	S	S
IBM	S	S	S
Motorola	S	S	S
Navistar	Ss	–	–
Warner Electric	S	m	–
Xerox	S	S	S

S = strong emphasis, m = moderate emphasis, – = not emphasized, s = strong emphasis on involving suppliers in the project team

use different combinations of methods and techniques the pattern of generic elements observed is as follows.

- The use of cross-functional teams to integrate the design of a product and other functions which have an impact on it.
- The use of computer integrated design and manufacturing methods, i.e. CAD/CAM/CAE, to support design integration through shared product and process models and databases.
- The use of a variety of analytical methods to optimize a product's design and its manufacturing and support processes, i.e. design of experiments, Taguchi methods, design for manufacturability and assembly (DFMA) and quality function deployment (QFD).

Other writers, like Voss and Russell (7), list integration mechanisms to implement CE. Some of these are direct contact, co-location, liaison role, cross-functional teams, secondment, role combination, permanent project team or cell, and matrix management. According to the findings of Voss and Russell, companies are searching for ways of developing integration to promote CE. Some companies used just one mechanism while others used more than one. Their findings also suggest that there are considerable difficulties in implementing some of the approaches. For example, cross-functional teams met with resistance from established functions, secondment sometimes left seconded engineers isolated from both their host and home functions, there was conflict between integrators and others in the organization, and permanent teams had very difficult relations with the rest of the organization.

Table 3.3 Findings of the work of V. Daniel Hunt

	Aerojet	AT&T	Boeing	Deere	Grumman	Hewlett-Packard	IBM	ITT	McDonnell Douglas	Northrop	Texas Instruments
Multidisciplinary teams	M	M	H	H	H	H	H	H	H	H	M
Design production process											
DfM	M	H	H	H	M	M	H	L	H	H	H
Administrative process	L	M	H	H	M	H	L	H	H	H	L
Computer Integration											
CAD		H	H	H	M		H		H	H	M
CAE			H	H			M		M	H	
CAM			H	H	M		M		M	M	M
CIM				H			M		H	M	
CALS			M	M				M			L
Quality tools and practice											
Statistical process control	H					H		H	M	H	M
Design of experiments	M	M				H		H	L		
Quality function deployment	H	H	H	M	M	H	M	H	H	H	L
Taguchi methods	H	H							L	H	L
Deming's methods						H			L		M

H = high, M = medium, L = low

Their conclusion was that care had to be exercised in selecting the appropriate integration mechanism and that the integration mechanisms described previously would be most appropriate when the following factors apply.

- Differentiation – where there are linkages between highly differentiated departments.
- Cross-functional requirements – where there is a need to take into account the requirements of other functions, particularly those downstream in the development process.
- Uncertainty – where there is a high level of uncertainty in the use, interpretation or content of the data.
- Intensity and frequency of two-way information flow – where there are major feedback requirements between departments or functions.
- Complexity – where there is a need to liaise between groups because of the complexity of the product or the task.

3.3   ORGANIZATION

Among the things most common to CE implementors is the use of organizational change. Organizational change is implied across three categories: organizational structure, personnel practices, and business practices and procedures. As applicable to organizational structure, change incumbent with the adoption and use of CE philosophy typically involves movement away from hierarchical organization towards a flat organizational structure. A deep hierarchical structure is particularly evident among companies whose internal organizations are aligned along functional boundaries. The use of multidiscipline teams whose members come from a number of functional units drives movement towards a more efficient and responsive organizational structure.

Personnel practices will also change. Direct reporting officials will normally be an individual's team leader who may, or may not be, from the same functional area of the company. Reward structures become far more heavily centred around team rather than individual performance. Emphasis on organic capability is renewed through leadership and a commitment to developing a more qualified workforce.

Changes to business practices and procedures may include establishing long term customer/supplier relationships and eliminating counter-productive competition policies (e.g. unnecessary over-specification of processes and procedures). Additionally, adoption of CE will require new accounting practices in accordance with a shift away from functional operation. For example, pursuit of time and cost savings in the development of new products requires the use of accounting procedures that reflect resources expanded by individual teams as opposed to those used by functional units. Additionally, decisions relative to the number, size and composition of multidiscipline teams, together with a determination of whether to support co-located versus distributed teams, must be made.

The recognition that organizational change is a primary mechanism in the adoption and use of CE philosophy does not, however, constitute a recommendation for reorganization. Rather, it is a recognition that the way one is organized goes a long way towards enabling effectiveness in the achievement and performance of CE. Organizational change does not, therefore, simply constitute change to organizational structure. This is only one facet of organizational change and a by-product of changes to personnel and business practices and procedures.

In some cases it may be sufficient to instruct product design and manufacturing engineering to rethink how they develop products. In others, it may be necessary to change the structure of the company. For example, Adam Opel made one director responsible for both product design and production engineering and then set up an advanced product study department. This consisted of both product designers and production engineers who work together. A small change such as this makes it easier for people to be ready for larger change.

Bold moves are needed to drive home the need for a new way of thinking and acting. Changing the way people think about product development is the key, and this must be reinforced to people at all times. As a prelude to its involvement in CE, Volkswagen disbanded its quality control department, which had a fine reputation throughout the industry for its control of tolerances. It was decided however, that the company was relying too much on the quality control department to solve problems in design and manufacture. Now, quality is considered a responsibility of all employees, so people formerly in quality control have been deployed in different departments where their experience is of greater value.

3.3.1 Organizational structures

Although the design of organizational structures can influence the patterns of behaviour and relationships in a firm, it is always difficult to predict what its effect will be because so much depends on the predispositions of the people concerned – how they perceive the structure in question and how they react to it. The design of the organizational structure can improve integration but difficulties lie in measuring the extent of improvement. For instance, it is pointless establishing an appropriate organizational structure if there is interpersonal conflict or animosity between different groups. The management therefore has to overcome these barriers and create an atmosphere of team approach. In fact, the reverse of this situation is also true. Team approach can have only limited impact towards integration if the structural design is deficient.

Lawrence and Lorsch [8] found that successful integration in the companies they studied was characterized by a great deal of open confrontation of issues by the members of different departments. This openness can only be based on a genuine understanding of other departments' outlooks, ways of working and problems.

Various organizational forms or structures can be identified, depending upon the degree of integration of the team leader/product manager within the functional organizational structure. This variation extends from functional managers having full authority over product development, through balanced authority in a matrix structure, to team leaders having full authority in a project team structure. These organizational forms have to be adopted with an eye to the particular problems they are designed to solve. No one structure will be appropriate for all products or all companies. One company studied by Lawrence and Lorsch changed its organizational form over the duration of the product development process in order to meet the company objectives better. Now let us examine the different types of structures used in product design and development and their inherent strengths and weaknesses.

(a) Functional structure

Figure 3.1 illustrates a typical functional structure of an engineering firm. The co-ordination of the departments is achieved formally by an executive committee or board of directors.

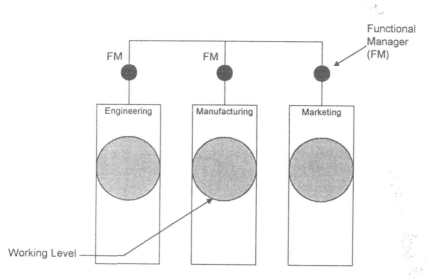

Fig. 3.1 A functional structure in which authority rests with the functional manager (adapted from Wheelwright and Clark [10]).

Child [9] lists the advantages related to the functional-orientated structure.

- It is economical on managerial manpower because it is simple.
- Co-ordination is left to top management rather than to integrating personnel such as departmental managers.
- When there are economies of scale or of concentrating resources in production (plant), in engineering (e.g. test or laboratory facilities), in distribution (e.g. warehousing) etc.; it is beneficial to group such facilities and the people operating them together into single functional departments.
- It provides clearly marked career paths for specialists and it makes it easier to hire and retain their services. They also have the satisfaction of working with colleagues who share similar interests.

Disadvantages of the functional-orientated structure
Problems arise when an organization diversifies its products, markets or services (diversification is very often a necessary condition for its continuous growth). If an institution is becoming larger, more diversified and subject to tight time constraints in adopting new products and services, then it is more than likely that a purely functional form will begin to break down under the strain.

However, the most serious disadvantages of the functionally-orientated structure is that the overall co-ordination of all the projects must be handled by the general manager. This has two disastrous effects.

1. The system does not afford the general manager the opportunity to delegate responsibility in relation to the product as a whole. This means that he or she is very much over-worked and therefore this can reflect in under-management of the organization.

   There is an important drawback in companies which have a functional structure. A conventional business structure assumes that business is a collection of separate disciplines which must somehow be welded into a business unit, for example, by having functions headed by an accountant, an engineer, a production manager and a sales manager, all reporting to a general manager or a chief executive. This means that no focusing of corporate opinion takes place except by the general manager and this leaves to one person the detailed co-ordination of all aspects of planning operations.

   A further disadvantage is that for the company organized on this basis overall co-ordination can really be vested only in the general manager. This is correct for those aspects of the business that a general manager should attend to, but immediately below there is **no one** in authority able to deal effectively and rapidly with project problems.
2. Furthermore, this type of structure provides scope for the different departmental heads to pass-off shortcomings as being due to the

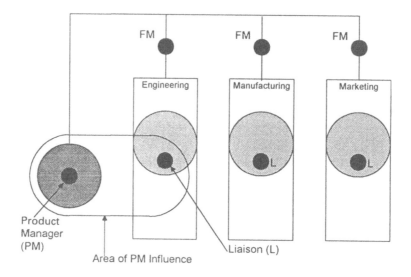

Fig. 3.2 A lightweight product manager form in which product managers/team leaders exist but have little authority in comparison with the functional managers (adapted from Wheelwright and Clark [10]).

failure of departments other than their own. This can lead to conflict, confusion and perhaps further deterioration in the degree of co-ordination.

(b) Product-orientated structure

A product structure becomes a more appropriate way to group activities when an organization produces two or more ranges of products, which are different in their technical make-up, production requirements and so on. There are two types of product-orientated structure, the lightweight and heavyweight product manager structure (Figures 3.2 and 3.3). These were derived by Wheelwright and Clark [10], where further details can be found. The advantages and disadvantages of this type of structure are shown below.

Advantages

- The groups are made up of personnel who have whatever training and experience necessary to develop the products.
- Its structure tends to create a more practical view of the work amongst the specialists involved.
- It is easier to assign the cost of research and development effort.
- The structure facilitates communication of information across the boundaries between subjects and disciplines.

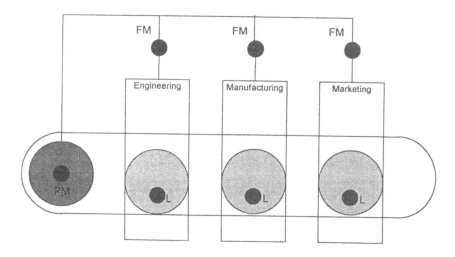

Fig. 3.3 The heavyweight product manager form provides the product manager with clear authority over the team members of the project (adapted from Wheelwright and Clark [10]).

Disadvantages

- This type of organization does not encourage engineers to keep abreast of the latest development in their fields.
- It creates an artificial subject boundary which makes recruitment more difficult.
- It reduces flexibility of labour.
- The equipment requirements are duplicated.

The most important advantage of this type of structure is that it provides the opportunity for the general manager to delegate authority and to bring about accountability for product development to lower levels in the structure. Thus, each product manager can be made responsible for each project being carried out by their personnel, whereas this is not possible in the functional structure. By delegating the responsibility to product managers there are further advantages to be gained, such as:

1. The product manager can be responsible for forecasting and controlling expenditure assigned to him or her.
2. The product manager can be responsible for planning and monitoring the progress of each project.

This way the general manager can relinquish detailed responsibility and thus devote more time to other aspects of the business, viz management of the organization. This can have further beneficial side effects.

- Product managers receive experience in management tasks and are likely to react enthusiastically to the enhancement of their responsibility; they have a better opportunity for achieving an understanding of project management than if the departments were function-orientated.

- Because of the wider nature of the work handled by the groups, the junior members of each of them are likely to receive far better training than in a function-orientated structure.

To illustrate some contrasting effects of organizing on a functional as opposed to a product basis consider the following example. Two manufacturing plants are making the same product for the same markets and using the same materials and technology. In the functional structure specialists focused sharply on their specialized functional goals and objectives. They identified closely with their counterparts in other plants and at divisional headquarters, rather than with the members of other functions in the plant or with common plant objectives. Their outlook was generally a short-term one, and the plant had a high degree of formality (job definitions, clear distinction between jobs) across all functions. In the other plant, the functional specialists seemed more aware of common product goals, with more of a product structure linking functions together. There was more variation in the time horizons adopted; for example, production managers concentrated on routine matters while quality control specialists were more concerned with longer-term problems. This differentiation in time horizon was encouraged by the way the product-orientated structure brought specialists together in problem-solving and led to a sensible specialization of effort.

Thus, in a functional plant, each department tended to worry more about its own daily progress. In a product-organized plant there were greater differences between functions in the extent to which organization was formalized.

(c) The lightweight product manager structure

The first product-orientated structure, the lightweight product manager structure, is shown in Figure 3.2. It is a structure in which there is a strong bias towards the functional organization and only a lightweight identification with the product – hence its name. The other functional specialists only contribute to the project on an intermittent and ad-hoc basis. They have no permanent role within the team, simply being called in by the product manager as and when needed.

Characteristics of the lightweight product manager structure

- no direct access to working people;
- less power in the organization;

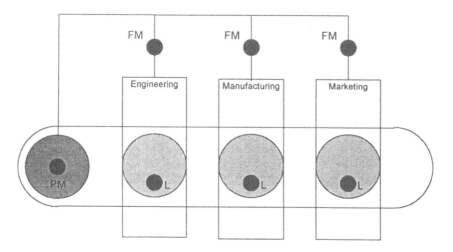

Fig. 3.4 In a balanced matrix structure the authority is shared between the functional managers and the product manager (adapted from Wheelwright and Clark [10]).

- has little influence outside product engineering and little influence in it;
- lack of direct market contact and concept responsibility.

The lightweight product manager's main purpose is to:

- co-ordinate;
- collect information on the status of work;
- help functional groups solve conflict;
- facilitate achievement of overall project objectives.

This can be summed up as a glorified progress chaser.

(d) The matrix structure

Here, a structure which is basically product-orientated is provided with a number of service groups which are of a specialist nature. The essence of this approach is to separate out those specialist functions which are not appropriate to segregate in the product groups. Figure 3.4 shows a typical matrix structure for an engineering company.

The case for matrix structure lies in the argument that it attempts to optimize two potentially conflicting benefits.

- It attempts to retain economic operation and development of technical capability associated with the functional grouping of common human resources.
- It attempts to co-ordinate those resources in a way which applies them effectively to different organizational outputs – products or programmes.

Child [9] lists the advantages and the disadvantages of the matrix structure, and these are as below.

Advantages

1. Helps to preserve flexibility.
2. Capacity to respond quickly and creatively to changes in a dynamic environment.
3. People are not 'wholly' members of a product or functional department – so easier for them to accept movement between teams and even departments if need arises.
4. Due to multiple reporting relationships and groupings of people it encourages open lines of communication within the organization as a whole.
5. Greater flexibility of people in the organization during uncertain conditions.
6. Claims to release a great deal of top management time from problems of operational co-ordination.
7. Because of the diversity of tasks and greater degree of authority undertaken by product managers, it helps their future development by broadening their horizons and expertise within the field of general management.

Disadvantages

1. The matrix structure attempts to formalize an already existing conflict between functional and product programme criteria. A third dimension of conflict may be formalized. This formalized conflict tends to generate conflicting objectives and accountabilities at a personal level, creating a highly charged political atmosphere with disputes about credit and blame and attempts to manipulate the balance of power.
2. The functional product managers will not always agree over priorities of resource allocation or over the time and cost allowed to functional activities.
3. The balance of power between the multiple authority structures is critical but delicate. It must be maintained if full benefits of the matrix are to be gained.
4. Employees experiencing threat to occupational identity are one source of stress.
5. Reporting to more than one superior can engender sources of anxiety and ambiguity.
6. Generally, greater administrative costs are incurred than in a more conventional structure.
7. The multiplications of hierarchies means an increase in managerial overheads.

8. The presence of conflict means that managerial time has to be devoted to its resolution.

Characteristics of the matrix structure

- The balance of power is split between the functional manager and product manager, there is therefore an unclear line of authority.
- Difficult to establish boundaries between functional responsibility, technical guidance and degree of design freedom.
- Often much time is wasted in working out accountability and responsibility issues.

(e) The heavyweight product manager structure

This is the second product-orientated organization structure. In it, the product manager carries full authority for the development of the product. The project team members fully participate in the development of the product, yet they remain attached to their functions. Thus the bias has shifted towards the product and away from the balanced matrix. Usually these type of managers are:

- more senior in the organization, or functional managers;
- have direct access to working level engineers;
- may lack formal authority, but exercise strong influence;
- are responsible for internal co-ordination, product planning and concept development;
- therefore function as a general manager of products.

(f) Project team structure

The project team structure consists of an autonomous project team, existing independently of the rest of the organization. The project team is assembled for a specific project under the auspices of the product manager. The team works exclusively and independently on the project until its objectives have been achieved. For the duration of its operation its members are permanent and full-time, having severed all connections with their functions. The team is thus temporary and will be disbanded when its project is completed. This is important as when the members return to their functional areas they transfer the knowledge, experience and skills they have gained from being a member of the team to their department. The product manager is usually slightly higher ranking than the functional manager. The former manager has complete freedom to allocate resources as and when necessary. This is the optimum approach for concurrent engineering (Figure 3.5).

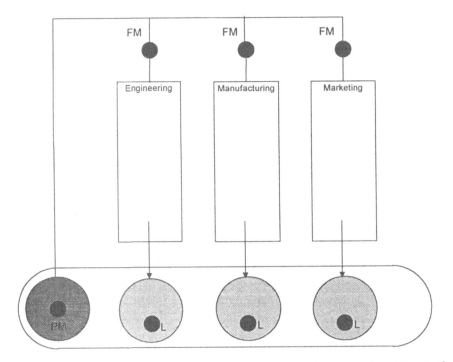

Fig. 3.5 The separate project team structure severs team members' ties to the functional department (adapted from Wheelwright and Clark [10]).

In this type of situation:

- the project manager works with a team of people;
- the team is selected from different functions;
- team members assume broader responsibilities;
- functional managers often retain responsibility for personnel development;
- product managers have greater influence on team members.

3.4 INTEGRATION MECHANISMS

Co-location is a technique for improving design and manufacturing integration. Locating the design and manufacturing engineering personnel in the same office helps informal consultation. This leads to problems being resolved earlier and more quickly. Many companies have found co-location improves relations between the two sets of engineers and thus improves time to market. One machine tool company reorganized its whole engineering department to place the design and production engineers in one open plan office. This greatly improved the informal communication between the engineers and resulted in improved designs being produced. The counter balance to this is the disciplinary attachments of the personnel concerned, which, despite their physical proximity, still prevented engineers talking to one another.

3.5 THE TEAM

Perhaps the most obvious and prevalent feature of CE is the use of interdisciplinary or interfunctional teams. More commonly referred to as 'multidiscipline teams', the primary mechanism for implementing CE philosophy centres around the assemblage of people with specialized skills, experience and perspectives on the product development process. Such teams are made up of capable individuals representing all departments of the company and all product life-cycle phases. The latter concern underscores the need to establish long-term relationships with, and facilitate intimate involvement in the process by, both customers and suppliers or subcontractors.

Successful use of teams is not new, nor is it unique to CE. Lockheed's use of highly motivated teams, called 'Skunk Works', was used effectively in the 1940s. 'Tiger Teams', 'Process Action Teams' and 'Integrated Product Development Teams' constitute only a small portion of representative examples where teamwork has been valued and practised. An important thing to recognize, however, is that use of multidiscipline teams does not necessarily constitute CE.

The team is the key lynch-pin of achieving time to market. A team philosophy is adopted for achieving integration. There are four elements to this team philosophy: responsibility, commitment, multifunctionality, and experience and proficiency [11]. First, the team must be given responsibility and be empowered for the introduction of the new product. It must be given the resources and authority to enable it to achieve its mission of rapid development. One way to provide authority is to have a senior manager as a team member, and possibly as chairperson. The team must take responsibility for managing the project, controlling activities and co-ordinating them. This can only be done through regular team meetings. Second, commitment of the team leader, team members and senior management is needed. The team should be headed by a manager who is committed to the philosophy of integration. The leader must be able to inspire commitment and enthusiasm from the team members. The team leader must also be an excellent manager in order to instil a team spirit within people and emphasize the co-operative and common tasks of the team to the members above their own specialist niggles and complaints.

Senior management must be committed to the integration philosophy and to team-working. They must give their backing to the project and allow team members the time to work in the team. This was illustrated in a farm machinery firm where the previous founder managing director did not allow design engineers onto the shop floor. This design-led approach resulted in the company's latest machine being 50% over cost. The newly appointed managing director has pursued an extremely proactive approach – encouraging and chairing meetings – to get the design and production personnel to talk to one another and overcome their former non-integrative approach. This had proved very successful and the new machine had a much more integrated design.

Thirdly, the team has to be multifunctional, drawn from the principal functions of the company. These are product design, process engineering, manufacturing, marketing, service, purchasing and selected vendors. Fourth, experience and proficiency of team members must be sufficient to enable the team to perform its role successfully. Each team member needs to be experienced enough and proficient at their individual functional tasks to ensure success. If team members are lacking in expertise or knowledge, for instance of the company's production facilities, training can be provided. During the selection process the importance of compatibility of members should be recognized. An existence of dominant characters can be a serious disadvantage to the spirit of the exercise. Therefore, a willingness to take part and commitment to the team is considered to be an inherent part of the selection process. The overall aim is to develop a well rounded and balanced team with the right blend of skills and experience. The size of the team must be manageable, not exceeding 12 members. This may mean breaking a product down into several teams, each being responsible for major components. Here, inter-team communication would become an issue for careful management.

The effectiveness of teams largely depends upon individual team members' qualifications, experience, technical competence and experience, the 'teamwork skills' and personalities of individual team members, overall team efficiency and the establishment of an environment for meaningful and productive interaction (often characterized as breaking down cultural and disciplinary boundaries). Undue emphasis should not be placed on the creation of incentives designed strictly to reward 'teamwork skills'. Such misguided attempts to value interpersonal skills on a par with technical competence and creativity most often nourish rivalry, politics and fear. Such skills should be cultivated and nurtured through empowerment, trust and encouragement. Leadership within individual teams and at every level of the corporation should replace merit-based systems in creating an environment for productive exchange and maturation of ideas.

The way in which teams are assembled and developed can go a long way towards establishing such an environment. Multidiscipline teams are built around a specific product or process for which they are responsible, which includes the notion of both accountability and commensurate authority. Any one individual may be assigned to a number of multidiscipline teams, either out of necessity or to leverage unique skills and experience.

However, experience in the use of multidiscipline teams suggests that continuity should be maintained in the people assigned to teams addressing similar developments efforts over time. This is not to say that active participation by all members of the team will necessarily occur throughout all life-cycle phases of the product. The extent to which continuity can be maintained is recognizably limited by additional factors. For example, an effective way of overcoming a seemingly insurmountable stall in progress is to replace the team leader.

Maintaining continuity in the make-up of multidiscipline teams over a number of similar development efforts enables individuals to develop mental linkages between the concepts, terminology and concerns of other team members (each representing different domain-specific perspectives) in relation to those peculiar to their own. Repeated exposure to similar development efforts also enables individual team members to recognize and apply knowledge to recurring situations.

Although CAD/CAM and a database are important aids in CE, electronic data interchange (EDI) is not a substitute for meetings. It might be thought that instead of holding actual meetings between members of the team, data could just be interchanged through a central database. However, this approach will turn the operations into over the wall engineering under a different name. The designer sitting at his terminal will be reluctant to send data until he has finished; the manufacturing engineers will be equally reluctant to send suppliers of machine tools incomplete data, and so sequential engineering is inevitable.

3.5.1 The team members and their style

The issues involved in the selection of a team are best explained by using an example of a company known as FG Ltd. The author was invited to assist this company to establish the team structure for a specific project.

Over the last few years the rationalization of FG's business in terms of reduced overheads and direct labour, and the adoption of advanced manufacturing technologies, new machines and product methods, resulted in the development of a leaner, meaner and generally more competitive productive unit. At this juncture, management were seeking a different perspective for improving profitability and remaining competitive in the market place. The emphasis changed from a concentration on manufacturing process to a concerted effort upon the **product-related** areas of the business – this seemed to be the obvious choice.

The measures taken initially, successful though they were, were primarily focused on the '**process**' side of the business. There was a move afoot, however, to shift the emphasis towards '**product**-related areas of the company. This should not compromise the efficiency improvements made to date.

Corporate managers identified the potential for further reducing direct costs through the redesign of existing products. This needed to be achieved without detriment to the value or quality of products, as received by FG's customers. If at all possible, the company should seek to simultaneously improve product features and performance.

The company, in essence, was moving to a higher plane of improvement – not just to improve process efficiency, but to actively seek to increase manufacturing effectiveness and therefore enhance overall corporate performance.

In addition to these reasons, managers were becoming increasingly concerned about various products in relation to product life cycle concept. It was recognized that a number of products were well into maturity and there were strong indications that decline was imminent in a number of cases. Thus the company was seeking the best way forward by setting up a team structure which could look at their existing product range and improve the product design. The need to remain competitive was seen as even more urgent due to the emergence of the European single market.

This company had acknowledged the fact that some companies employ full-time design specialists, but experience, however, shows that this is not usually satisfactory.

- Departments such as design, buying, finance, production and marketing may not be fully consulted during this exercise, therefore personnel in these departments may feel undermined. This of course may lead to resistance or even rejection of changes proposed to the product.
- Selling the proposed changes throughout the organization becomes difficult.
- There is loss of potential intangible benefits for the team members.

Therefore, it is often better to have part-time members from a number of departments with diverse experiences who may well be involved with implementing product and process modifications at the implementation stage. This develops a sense of ownership through the organization, invaluable when attempting to put changes into practice.

In the case of FG Ltd the team to work on the first project was selected only after lengthy discussion between the author and the senior management of the company. A total of 12 members were chosen from a variety of functions. The criteria for selection included age, qualifications, specialist knowledge of product, knowledge of manufacturing methods, experience and functional area. During the selection process the importance of compatibility of members was recognized. An existence of dominant characters could be a serious disadvantage to the spirit of the exercise. Therefore, a willingness to take part and commitment to the group was considered to be an inherent part of the selection process. The overall aim was to develop a well rounded and balanced group with the right blend of skills and experience. Based on this the group chosen were as follows: stock controller, project co-ordinator, design draughtsman, production superintendent, undergraduate business students, foreman, facilities engineer, an accountant, work study engineer, prototype technician, marketing executive and head of planning. The group was multifunctional and had a cross-section of experience. It was further split into three teams. Each team concentrated on specific aspects of the product under examination.

The work of the group was co-ordinated, facilitated and directed by the author of this chapter, who was assisted by the General Manager at FG Ltd.

This case study shows some of the issues which need to be considered. The choice of the right mix of team members is crucial to the effective functioning of the team. It is management's responsibility to select team members to build an effective team.

The following section details some of the characteristics of team members and their contrasting opposites. This can be used to select team members to ensure a balance of characteristics appropriate for the tasks and goals the team has to meet. Teams have to be chosen to reflect the demands and requirements placed upon them, for their particular tasks – not every task can be successfully tackled by the same team! Emphasis must also be given to the team's ability to separate their task performance from their process performance. This is critical to the identification of lessons learnt and further implementation into the organization.

(a) Talkers and Reflectors

Talkers
- generate and discuss ideas
- brainstorm.

Reflectors
- like to ponder
- not eager to share until sure
- think before speaking.

(b) Planners and Adaptors

Planners
- think in an orderly fashion.

Adaptive
- are spontaneous
- go-with-the-flow style.

(c) Detail Specialists and Visionaries

Detail specialists	● give and receive factual data when communicating.
Visionaries or possibility thinkers	● gather data and quickly establish interactions and patterns.

(d) Thinkers and Feelers

Thinkers	● consider pros and cons of facts.
Feelers	● subjective or value-based decision making
	● strive for cooperation and harmonious working conditions.

3.5.2   The two-team approach

The other important point is to separate the strategic and operational elements of the team. This can be done by adopting the two-team approach, as detailed by Riedel [12]. The first, or strategic, team was convened to consider the strategic issues of product design.

This team would be responsible for setting and controlling budgets and setting time scales. It would also review the progress of designs currently in development and correlate these with marketing. This team would consist of senior managers of the relevant functional areas, including company directors in some cases. The second, or operational, team was charged with the day-to-day management of the process of introduction of a new product. It would thus be concerned with literally the nuts and bolts of the design and development. The team would typically consist of design and production engineering personnel plus factory representatives. It would consider the manufacturing problems of the design and how to resolve them. This two-track approach to the management of the introduction of new products was found to be more successful for the companies using it than for firms not following the approach [12].

The strategic team also has to take responsibility for forming the operational team, maintaining its dynamism and disbanding it at the end of the project. For the next project a new team should be assembled from different personnel, to provide an element of challenge and prevent team members from becoming complacent. This also serves to distribute CE skills and knowledge throughout the organization. It also percolates the knowledge gained from implementing CE around the functional areas of the organization.

3.5.3   Intensity of integration

Managing the intensity of integration is one of the trickiest aspects. It has been best discussed by Takeuchi and Nonaka [13]. The rhythm of the development team's activity varies across different development phases. It is most intense in the early stages and tapers off toward the end. This requires the careful management of rhythm throughout the development process. Secondly, if a bottleneck develops during development there is a rise in activity (people searching for solutions) but the total process does not come to a halt. Thirdly, integration requires extensive interaction among project team members but also with suppliers. This interaction increases speed and flexibility. Fourthly, the previous 'hard' benefits are matched by 'soft' ones: 'it enhances shared responsibility and co-operation, stimulates involvement and commitment, sharpens a problem-solving focus, encourages initiative taking, develops diversified skills, and heightens sensitivity toward market conditions' [13]. De-merits result from having to manage a more intensive process: 'problems include communicating with the entire project team, maintaining close contact with suppliers, preparing several contingency plans and hand-ling surprises' [13]. The approach also creates more tension and conflict within the project team.

3.5.4 Leadership

Adler *et al.* [14] consider the role of senior management. They argue that 'extensive and systematic participation in the pre-project phase of a product development effort is probably the single most important area where senior management can put its involvement in technology into operational practice'. This need for early involvement was shown to be beneficial in a gas fire firm, where the early start given by senior management resulted in the fire being introduced only eight months later without delays. However, once the design process is underway senior management need to stand back and not interfere directly. This hap-pened in a food preparation machine company, where management interference slowed down the design process. The marketing director presenting new specifications to the design team while the product was in the middle of being designed, thus causing redesign delays.

In the context of CE, there are four categories of leadership qualities that are relevant at all levels of an organization and which become particularly important for informal leadership situations. First, effective leaders demonstrate professional competence which includes a thorough knowledge of the job and the ability to apprentice subordinates towards mastery of the same skills and knowledge. In addition to being capable as a specialist, leaders develop the capacity to see the 'big picture', recognizing the extent and limitations of specialized knowledge within the context of the overall problem. More fundamental skills, such as the ability to communicate both orally and by the written word are often cited as vital to effective leadership. An additional connotation of professional competence might best be characterized as 'situational awareness' or the ability to recognize and appropriately respond to different situations.

The second category is what might be called a leadership perspective. An effective leadership perspective views position as an obligation to subordinates, not as privilege and honour for which one is uniquely deserving. An effective leader recognizes that respect is earned, not awarded. Leadership perspective, as opposed to a strictly supervisory management perspective, is an attitude demonstrated through action rather than words, such as taking an active interest in people and in what they do, valuing individual ideas and opinions, and creating opportunities for others. Leadership perspective is the view acquired when leaders do not allow themselves to become tied to the inside of their own offices.

Strength of character is the third category of leadership attributes. Among the qualities commonly cited for effective leaders are reliability, courage, dedication, integrity, determination and self-discipline.

Fourth is a category called inspirational qualities. This is the ability to capture the imagination of others, and to inspire them to achieve what they would otherwise have been incapable of achieving. Ultimately, the true mark of an effective leader is one who is able to gain and maintain the respect and confidence of his or her subordinates.

The ultimate success of CE efforts will be determined in a large way by effective leadership at all levels of management, besides state-of-the-art technology, formal methods or even the assemblage of highly skilled multidiscipline team members.

3.6 TIME

Managing time is, of course, the fulcrum of concurrent engineering. Some companies use milestones as one way of managing time. The project leader of the project team will hold meetings at the following milestones. Before and after prototype construction, and before and after production machines were produced. The first production meeting may, or may not, be held, depending upon the outcome of the second prototype meeting. Six months after product introduction another milestone meeting is held to check that everything was OK. For each project a project review meeting would be held monthly, six weekly or more often as necessary. These meetings would review progress and set deadlines. This approach fits neatly into the two-team approach to product development, with the strategic team having the responsibility for managing time. The timeliness of these operations has a logistical aspect, which is considered next.

3.7 LOGISTICS

Logistics are an important part of the integration equation. The capability to manage one's internal operations (design and manufacturing) and external operations (subcontractors) is crucial for achieving time to market. This capability is illustrated by a major machine tool manufacturer visited by the author. The company had successfully fended off competition from the Japanese and Koreans by shifting its operations to 'buy-in-parts' rather than manufacture them in house. Their latest machine only had 73 in-house manufactured parts, the rest were bought-out. This reduced the lead times on the last three new machines the company had introduced.

The improvement had come about due to three factors. First, by basing new designs on current machines and thus producing families of machines. The first machine thus introduced was completed in seven months. The company had never done a new design so fast. A further two machines were also introduced in a similar time. The second factor was the project team management. A problem with this approach was that the company did not have enough people available for teams. The third factor was organizing suppliers and internal manufacture to deliver parts when needed, and quicker than before. It was this logistics capability of the firm that had been key to its competitive edge.

3.8 CONCLUSION

This chapter has provided an overview of how companies can and have organized and managed concurrent engineering to introduce new products more effectively. The live examples given demonstrate that integration is achievable and does improve competitive performance – both to stave off competitors and to beat them. This chapter considered achieving integration through concurrent engineering, integration methods and tools, organization, organizational structures, integration mechanisms, management issues of teams, team selection (including an example), two-team methods, intensity, leadership, targets, time, logistics and data and information. It shows that each of these approaches needs to be adapted to the company context and the products to be developed. Careful thought has to be used to select which methods and techniques to use. It is management's responsibility to make these choices and to manage the use of concurrent engineering. Once the decision has been made to adopt CE, management must learn from experience, smooth over the inevitable bumps, change tack, but maintain their eyes on the prize of achieving set targets for gaining competitive edge.

REFERENCES

1 Pawar, K S (1985) An investigation of the nature of the working relationship between product design and production functions in manufacturing companies PhD Thesis, University of Aston

2 Gillen, D J (1991) Expanding Knowledge and Converging Functions *Concurrent Engineering*, **Jan/Feb**

3 Hartley, J (1990) *Simultaneous Engineering* Department of Trade and Industry, London

4 Ettlie, J E and Stoll, W (1990) *Managing the Design–Manufacturing Process* McGraw-Hill, New York

5 Trygg, L (1992) *Simultaneous Engineering A Movement or an Activity of the Few?* International Product Development Management Conference on New Approaches to Development and Engineering, Brussels, 18–19 May 1992, European Institute for Advanced Studies in Management, Brussels

6 Hunt, V D (1991) *Enterprise Integration Sourcebook* Academic Press, San Diego

7 Voss, C A and Russell, V (1991) Implementation Issues in Simultaneous Engineering *International Journal of Technology Management*, **6** (3/4)

8 Lawrence, P R and Lorsch, J (1967) *Organization and Environment* Graduate School of Business, Harvard University, USA

9 Child, J C (1977) *Organization, a Guide to Problems and Practice* Harper and Row, London

10 Wheelwright, S C and Clark, K B (1992) *Revolutionizing Product Development* Free Press, New York

11 Bower, J L and Hout, T M (1988) Fast-Cycle Capability for Competitive Power *Harvard Business Review*, **66** (6) Nov–Dec, 110–18

12 Riedel, J C K H (1991) Case Studies of Product Design Management in Mechanical Engineering, Working Paper, Economics Division, Wolverhampton Business School, Wolverhampton Polytechnic

13 Takeuchi, H and Nonaka, I (1986) The New New Product Development Game *Harvard Business Review*, **64** (1) Jan–Feb, 137–46

14 Adler, P S , Riggs, H E and Wheelwright, S C (1989) Product Development Know-How Trading Tactics for Strategy *Sloan Management Review*, **31** (1), 7–17

CHAPTER 4

Design maturity

C. O'Brien and S. J. Smith

Launching a high quality saleable product before the competition is vital for success in the current market for electronics goods. Concurrent engineering (CE) is an approach designed to fulfil this objective.

The electronics industries have led CE development. In any company there is a need to select the most appropriate combination of tools and methods to meet its strategic, tactical and operational objectives within the CE framework. Some tools still need to be developed to improve the CE performance objectives of the electronics industries. Examples include rapid costing and design maturity assessment tools.

4.1   THE PRESSURE ON PROFITS

Combining teams, tools and formal methods allows the fullest use of CE. This requires collaboration and co-ordination. Specifically, it is the flow information generated through the processes of design and development that needs co-ordination. One of the major benefits of using CE is the reduction in product development time. This is accomplished through the simultaneous development of various aspects of the product, shown graphically in Figure 4.1. Development time and cost are becoming crucial in all the engineering industries, but are particularly serious for the electronics industries, whose profits have been squeezed the most over the last decade. Figure 4.2 [1] shows the reduction in product life span (PLS) across all engineering industries.

The reduction in profits requires a solution. A reduction in the cost and time involved in product development is one solution, CE being the method required to attain this goal.

4.2   DESIGN STRUCTURES

Closer examination of the concurrent engineering development time diagram, Figure 4.3, shows that the start of each activity is phased. This

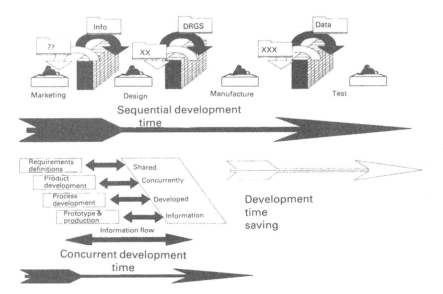

Fig. 4.1 Product development for sequential and concurrent engineering.

can be clearly seen in electronics design. The following diagrams show that sequential activities precede parallel design activities, and parallel activities contain within them sequential activities. The diagrams are IDEF0 charts of the electronics industry, with particular focus on the design process. Figure 4.4 shows the whole design process and demonstrates that hardware and software design are simultaneous, but only after the determination of the sub-system specifications. Design is itself a

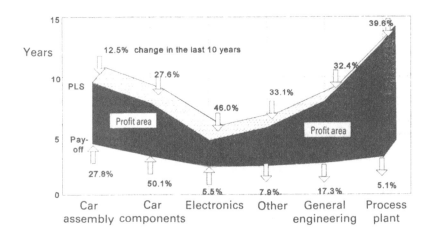

Fig. 4.2 A comparison of product life span and pay-off period (to 1991).

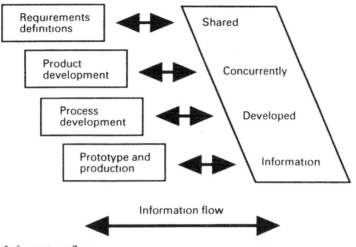

Fig. 4.3 Information flow

parallel and sequential process depending upon its characteristics. Figure 4.5 illustrates that hardware design is also a parallel process in that ASIC, PCB and other designs like power supply, can all be done simultaneously. This does not however mean that each activity is also parallel, and examination of PCB design in Figure 4.6 shows that this is still sequential. Should you wish to know more about this, IDEF0 Charts and details of the Electronics Design process can be found in reference [2].

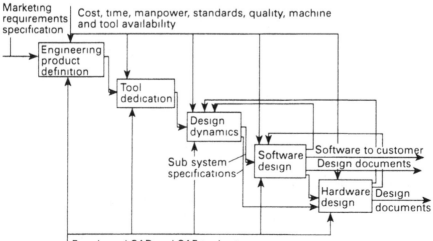

Fig. 4.4 Design – A3

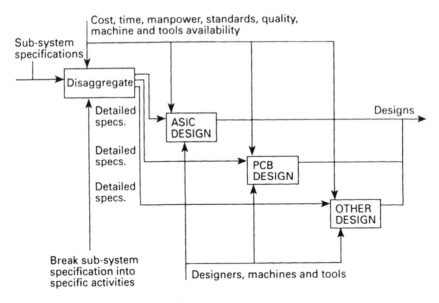

Fig. 4.5 Hardware design – A35.

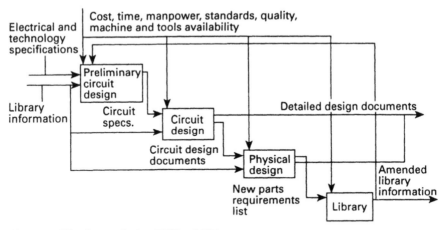

Fig. 4.6 Hardware design PCB – A353.

4.3 DEPENDENCY AND DELAY

Each area of design has a cycle of activity level over time. Figure 4.7 displays this using 'haystacks'. CE has brought these haystacks closer together so they now overlap. Each activity still has an independent phase before an area in which everything in which it participates has a dependent characteristic. After a certain point in time everything done in the activity will affect the activities that follow.

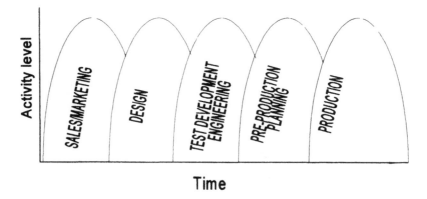

Fig. 4.7 Functional relationships.

The point at which other activities become dependent on the current activity is when they release information that they are dependent on for their own tasks. Figure 4.8 shows the independent and dependent zones. In a CE environment, as in a traditional environment, a change after the release of information to the following activities in the sequence increases the overall workload and delays the product launch. Such delay extends the duration of the following activities even if it does not delay the activity that has made the change. Figure 4.9 shows this. The release of information to more activities needs correction in more areas than before; premature information release can cause greater delay and more work in a CE environment than it would in a traditional design environment.

Alternatively, if a design is ready for distribution to the other dependent activities but its despatch is delayed, then the whole process is delayed and again so is the product launch. Figure 4.10 shows this effect.

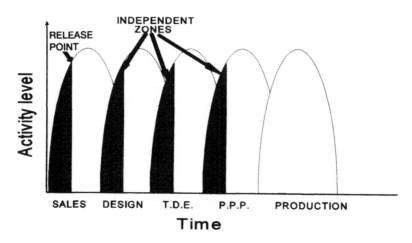

Fig. 4.8 Dependent and independent zones.

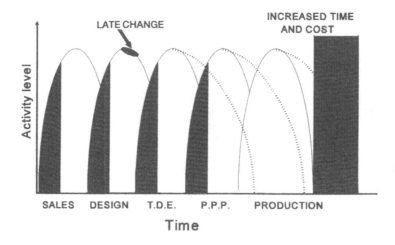

Fig. 4.9 Result of change after release

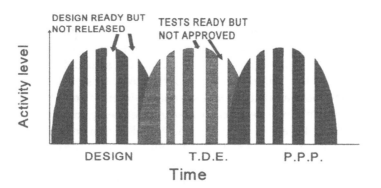

Fig. 4.10 Delays to good design by phase reviews and checking

4.4 THE EFFECTS OF ERROR AND DELAY ON PROFITS

Error and delay have a direct effect on profit and in the electronics industries the effect can be large. Jim Goodman of GPT says that electronics projects can be divided into two main groups, very large capital projects and smaller capital and retail type projects. The effect of late delivery on the customer's loyalty depends on whether the project is large or small. If a large bespoke project is over-running the customer may want to wait for delivery. If it is a small investment however, and an alternative source becomes available, the potential customer is likely to select the alternative rather than wait, so that they can equip themselves immediately.

With most projects the current estimate of the effect on profit of late delivery is that a design over-run by 100% halves the profit. If the product under development is in the smaller group, such an over-run may mean it never gets into the market place as the competition will have filled its niche. Such a product is not only profitless but becomes a financial burden on the company. Current estimates are that most late running projects will come under careful management before reaching 100% over-run (i.e. replacement of the design manager).

Goodman estimates that if all projects ran to schedule, profits would increase by 100%. In a CE environment the increase could be even greater.

4.5 WHAT IS MATURITY?

Designs may be considered to evolve towards stability or maturity, but when is a design mature enough to begin to release design details to downstream activities?

A.A.J. Willitt's [3] description of maturity is:

> Maturity makes possible a 'linear' design process, each stage flows sequentially – no back tracking. Non-maturity results in an 'orbital' process, a circular sequence of repetition, the total time taken being several times that for the linear process.

Using this definition of design maturity, maturity in a CE environment is that moment when a design is complete enough to allow the release of details to downstream activities, knowing that further design within the current activity will not lead to redesign within any of the downstream activities.

Knowledge of the moment of stability will reduce the risk of releasing an immature detail to a following activity. The delays caused by the need to check, through other methods, the validity of the design before its release would be removed as well. This leads to the question of which co-ordination methods for the management of the design process are available and what use can be made of them to guarantee the security of the process in a CE environment.

4.6 METHODS OF DESIGN CONTROL

The methods currently used for the product development team control process are:

- configuration management;
- design reviews;
- project management and its adaptation, total time management;
- checklists.

All the co-ordination methods have problems.

Configuration management involves ensuring that design is phased and that at the end of each 'phase' all the outputs from the various areas of design match, before starting the next phase of design. For example, checking to ensure that all the design teams are using 6 volts peak-to-peak as the relevant inputs and outputs for a board, not a mix of RMS and peak-to-peak. This may lead to the delay of a product introduction because of the time taken checking the design compatibility among design areas, before the release of information from all currently simultaneous design areas to the next phase of design. The delay of information delivery from any area will stop the process until its completion and approval.

Design reviews by individual managers can be protracted as they check every detail of a design before its release to ensure that there is no need for late changes. This method of control depends on the manager's confidence level in the individuals within the design team that they control.

Design review by committees which can only meet when all members are available may delay release by however long that interval is. Project management looks like a successful method for adaptation to manage design. Project management is a method used successfully in an environment in which there is a knowledge of nearly all or all parameters, a deterministic environment. Design, however, is about the discovery of those parameters used by project management to manage. This can lead to conflicts caused by the mismatch between management style and working style. To quote Dr. J. Warschat of the Fraunhofer Institute:

'In practice . . . the following contradictions occur:

- Chaos has a positive effect on the creation of order, but should be reduced from the beginning.
- Processes cannot be determined, but it should be possible to control them in a deterministic way.
- At the beginning, the aims are unclear, but they should be known exactly from the start.'

In applying total time management, design could suffer from the same effect, as design is a stochastic process which eventually leads to a deterministic plan.

Using a specifications checklist to determine the maturity of a design will ensure the released design is mature. However, it will also be complete. Large proportions of design involve the discovery of the engineering specifications that overlap the customer's requirements. Design is the translation of customer and marketing requirements into engineering requirements. A large part of the design process itself is the construction engineering specifications. Imposing specifications on design before the start of a project inhibits the design process. This can result in designs which are not the best being produced. The designs will have to conform to a set of guidelines that may not be in the interest of the final customer.

Specification production for use later in the design process is itself an act of design and needs careful consideration, like the rest of the design process, to ensure the specifications generated meet the requirements of the end user. Figure 4.4 illustrates this, where parallel design started after the writing of the sub-system specifications. The maturity of the specifications lists and other checklists used to monitor the compliance of the remainder of the design process with the customer's requirements, needs assessment to ensure that these will not change after their release downstream. A change in the specifications would have as serious a consequence on the final delivery date of the product as any other loop within the design system. Specification changes are often cited as the main reason for late delivery of a newly designed product.

In any design area it is necessary to be able to assess when the design is mature enough for release. It is necessary to understand what is to be co-ordinated and how this will be managed. In a concurrent engineering environment the objective is to share information from specialist areas as soon as possible with those activities that will help in the conversion of a design into a product. By sharing common information, some processes traditionally performed sequentially can be accomplished in parallel. However, sharing information that is immature or incorrect will **drastically increase** (by the number of areas now receiving incorrect information compared with the number in a traditional design environment) the overall design effort and lead to 'concurrent chaos'.

The need to understand the co-ordination process is becoming a recognized area of research. The MIT Sloan School of Management has set up a centre for co-ordination science. Thomas Malone [4] recently produced a technical report entitled *Towards an Interdisciplinary Theory of Co-ordination*. The narrow definition of co-ordination given by MIT is:

> **Co-ordination is the act of managing inter-dependencies between activities.** Co-ordination involves **actors** performing **interdependent activities** that achieve **goals**, and its analysis includes goal decomposition, resource allocation, synchronization, group decision making, communication and the perception of common objectives.

Within design the activities may be classified as:

- **pooled**, where the activities share common information but are independent in action;
- **sequential**, where they depend on other activities decisions prior to their initiation;
- **reciprocal**, when each activity needs data/information from another in a back and forth process.

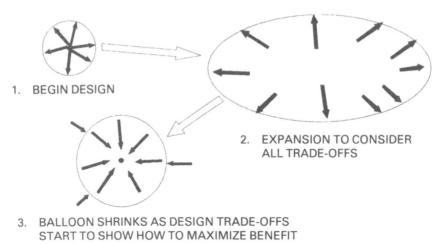

1. BEGIN DESIGN

2. EXPANSION TO CONSIDER
    ALL TRADE-OFFS

3. BALLOON SHRINKS AS DESIGN TRADE-OFFS
    START TO SHOW HOW TO MAXIMIZE BENEFIT

Fig. 4.11 Cost model balloon.

To manage the design processes in a concurrent engineering environment more effectively, the maturity of the individual components of
design need to be known all the time. When known, co-ordinating the
start of the pooled and reciprocal activities that are sequential to the
initiating activities can be done with confidence. Noble and Tanchoco [5]
discussed the balloon model of costs in designing a product which Figure
4.11 illustrates. After some time the designers have learned enough for an
idea of the final product to crystallize in their minds. The designer(s) can
then focus on producing the design specifications and drawings. The
process initially involves large changes in what is included and deleted
from the design model. As the design progresses the changes become less
dynamic in nature. However, after some progress, detection of errors in
the original theory for the model will lead to more large changes until
finally an agreeable solution to the design problem is found. The
designer(s) can then settle down to the detailed design. Figure 4.12
provides graphical illustrations of some possible patterns of design
change dynamics. Today most electronics design is now carried out with
the aid of computers. This in turn has led to the development of
electronics data management tools (EDM). These tools carry out a
multiplicity of tasks and free the designer of much of the tedium that used
to be necessary as part of the design task. Examples can include automatic
translation of design parameters into different formats for testing in
models and the tracking of version numbers of designs.

   Using the capabilities of EDM tools it should be possible to determine
which patterns fit particular designers or design groups and to then value
the current status of the designs. Using this information it may be
possible to quantify the size, importance and frequency of the changes
made to the design throughout its life.

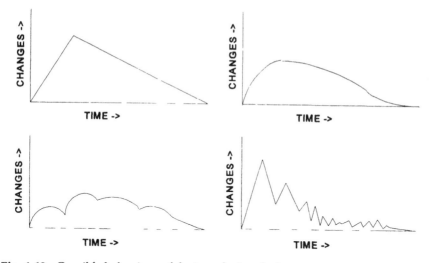

Fig. 4.12 Possible behaviour of designs during design.

4.7 DESIGN MATURITY ASSESSMENT

A tool that assesses the maturity of designs (that is, the risk involved in releasing a design) under consideration for release in a CE environment would be of great benefit to design managers, providing a higher degree of confidence in the safety of release.

It may be possible to provide probabilities of maturity for any particular design activity at any point in its life cycle, for example 25% confidence that the current design is stable, 98% confidence in the maturity of the specification data, and so on. This would speed the release of designs of sufficient maturity and prevent the release of designs that have not reached the required maturity.

With sufficient data it may be possible to determine the earliest times at which it is economically feasible to take the risk of releasing design details from initiating activities to the other areas of design that are sequential. It may also be possible to detect when a design group is in difficulty and needs specialist help.

4.7.1 Methods of measurement

It may be possible to define the degree of design maturity by one of several methods:

1. Determination of the number of changes made to a design over time.
2. Determination of the size of the changes made to designs by designers.
3. A combination of methods 1 and 2. This is probably the best option.

4.7.2   Data collection

The method of collection of the data that measures the number and magnitude of changes listed above is important. The easiest and perhaps the least reliable method of data collection would be to ask designers to fill in a worksheet on which they state the number and magnitudes of the changes they make. The flaw in the method is that as a project deadline approaches, irrespective of the number and size of changes made, a designer might state that the changes made are minor, so that his or her design will be passed.

4.7.3   Change frequency

It is possible to determine the frequency of changes made to a design automatically by monitoring the CAD files the designer has created. Each change generates and stores a new version number of the design file so the designer can backtrack if required. This allows the measurement of the number of changes per unit time.

4.7.4   Change magnitude

It is not considered possible, however, to determine the magnitude of change from examination of the change notes, file numbers, dates or times.

This information could be collected by using historical design files and asking designers to look at the CAD file progression (the changes from version 1 to version 2 and so on) and score the changes made for magnitude (say between 1 and 100).

4.7.5   Risk evaluation

Using the same design files, assessment of the risk levels involved in design data release can be determined. The number and magnitudes of changes made to the designs after their release can be tackled by scoring the CAD file changes (magnitude and frequency) after design release, and the magnitude and frequency of engineering change notes received after design release. This can be measured in exactly the same way as the pre-release version files.

Application of this information may assist design managers in the maturity assessment decision and may enable them to release designs earlier with a higher degree of confidence in the safety of release.

It may be possible to produce probabilities of stability for any particular design activity at any point during its life cycle, for example 25% that the current design is stable and 98% confidence in the specification data.

With time and learning it may become possible to determine the earliest times at which it is economically feasible to take the risk of releasing initiating design details to other areas of design that are sequential. For example, to know that it is possible to release specifications A–F at 80% maturity and G–Z at 97%.

A further extension to this may be the ability to detect difficulties that may need even more specialist help. It could be that if the percentage maturity varies widely from day-to-day, or has stopped increasing, that specialists need to be placed on the design team. With a large enough volume of material general trends in design for individual designers and groups of designers may become apparent, allowing even better assessment and letting the tool to be used for project forecasting pre- and in-process.

4.8 SUMMARY

CE involves design teams whose activities may be:

- pooled
- reciprocal
- sequential.

When the activity is sequential the delivery of design information from the previous activity is critical. The information delivered must be mature. Co-ordinating the design activity so that the information is mature relies on processes that may not be reliable or efficient. Using EDM tools may make it possible to monitor design progress to determine when information is stable enough to release. Information may then be released in an optimum sequence to maximize pooled and reciprocal design activities. Optimizing design release could double profits.

REFERENCES

1. Warschat, J. (1991) *Just in Time Product Management*. Proceedings of the International Conference on Computer Integrated Manufacturing ICCIM '91, Singapore, 2–4 October 1991, pp 185–8.
2. O'Brien, C. and Smith, S.J. (1992) Productivity model for electronics design. *Journal of Electronics Manufacturing*, **2**, 71–88.
3. Willitt, A.A.J. (1985) *Thinking About Design and Quality*. Colloquium on Design Equals Quality: True or False? London, 7 March 1985. IEE Colloquium (digest) N1985/22.
4. Malone, T.W. (1991) *Towards an Interdisciplinary Theory of Coordination*. (Technical report). In the proceedings of the 1st International Conference on Enterprise Integration Modelling Technologies. ICEMIT, Hilton Hotel, South Carolina, USA, 1991 June 8–12.
5. Noble, J.S. and Tanchoco, J.M.A. (1990) Concurrent Design and economic justification in developing a product. *International Journal of Production Research*, **28** (7), 1225–38.

PART TWO

Essential Techniques for Concurrent Engineering

CHAPTER 5

Quality function deployment: an overview

U. Menon, P. J. O'Grady, J. Z. Gu and R. E. Young

5.1 AUTHORS' NOTE

Quality function deployment (QFD) is an important tool to facilitate multi-functional planning and communication in a concurrent engineering product development environment. It provides a structured framework to translate the 'voice of the customer' into the actions and resource commitments needed to meet customer expectations. User experiences confirm that QFD can facilitate the following:

- reduce product development cycle time;
- improve customer satisfaction;
- increase competitiveness.

This chapter presents an overview of QFD in terms of its evolution and current approaches to its application in numerous industries.

5.2 INTRODUCTION

The global market place for engineering products has become highly competitive. The market leaders for some strategically important industries (consumer electronics, steel and automobiles) have shifted to Pacific rim nations. Notwithstanding the importance of high product quality, it is recognized that producing defect-free products alone is no longer sufficient to guarantee customer satisfaction and survival for companies in highly competitive product markets.

QFD, which originated in 1972 at the Kobe Shipyard in Japan, gained widespread adoption in Japanese industry, and was subsequently adopted in the Western hemisphere during the mid-1980s onwards. A comprehensive discussion of the Japanese perspective on QFD can be found in Akao [1]. QFD applications in North America are outlined in

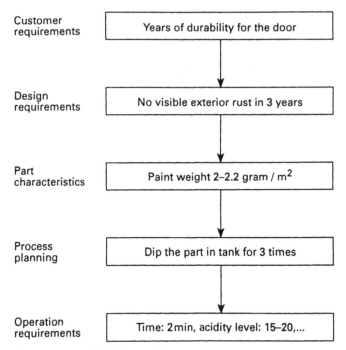

Fig. 5.1 Four phases of QFD translations.

Bossert [2], Hauser and Clausing [3] and Sullivan [4], with European experiences outlined in Langdon [5] and Dale and Best [6]. We are now at the mature stage of QFD implementation and can sustain the claim that QFD is an effective tool for systematic capture of customer needs and addressing those needs in a structured manner within multi-functional product development teams. QFD has been successfully applied to a variety of industries, notably for automobiles, aero-space, copiers, defense, consumer goods, electronics, textiles and computer software.

5.3   WHAT IS QFD?

QFD is a structured planning tool of concurrent engineering which can be used to influence the incorporation of product attributes which are in accord with customer expectations. This is done by mapping the customer requirements into specific design features (and eventually into manufacturing processes) through one or more matrices of 'expectations and fulfilment options'. QFD is used as a systematic approach to both identify and prioritize customer requirements, and to translate these requirements into product and process specifications [7]. Hence, QFD is a major tool for some aspects of contemporary implementations of concurrent engineering [8].

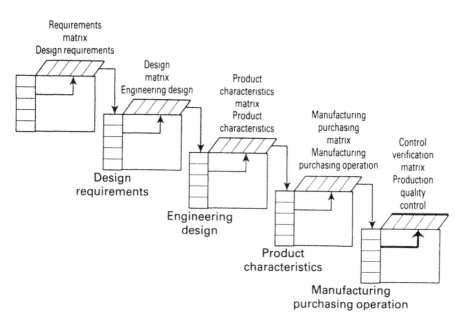

Fig. 5.2 The translation process of QFD.

Consider the classic QFD example, the design of a car door [3]. The QFD procedure starts with identifying the key customer requirements. If one of the requirements is that there should be three years of durability for the door, then this can be translated into the design requirement: 'no visible exterior rust in three years in an open air environment', as depicted in Figure 5.1. This design requirement can then be translated into the part characteristic 'paint weight 2–2.2 g/m^2', to fulfil that design requirement. The part characteristic is then translated into the process plan 'dip the part into the dip tank three times' which will produce the part with the required characteristic. The process plan is translated into the specific operations and conditions 'dipping time 2.0 minutes minimum, acidity level 15–20, and temperature 50° C' [4]. This procedure is usually carried out by using a series of translation matrices (Figure 5.2). Thus, QFD moves from the customer requirements (three years of durability for the door), through design requirements (no visible exterior rust in three years), part characteristics (paint weight), and process planning (dip the part in the tank three times), to the detailed operation requirements (dipping time 2.0 minutes minimum). Therefore, if QFD is implemented properly, customer requirements can influence the functional areas of the company that are often isolated from adequately knowing and reacting to the needs of their customers.

5.4   THE HISTORY OF QFD

The QFD methodology was developed in the late 1960s and early 1970s in Japan by Dr Shigeru Mizano of the Tokyo Institute of Technology. In 1972, Dr Mizano used 'quality tables' to help support planning, and these later evolved into QFD. Hauser and Clausing [3] state that QFD originated in 1972 at Mitsubishi's Kobe Shipyard site. Toyota Autobody began using QFD in 1977 and has experienced significant benefits, including a 40% reduction in the development cost for a new model and 50% reduction in development time. QFD has been used successfully by Japanese manufacturers of consumer electronics, home appliances, clothing, integrated circuits, synthetic rubber, construction equipment, and aircraft engines [9]. QFD was introduced into USA in 1983, essentially through the Ford Motor Company and the Cambridge Corporation. Dr Don Clausing was one of the pioneers in introducing the methodology of QFD to engineers at Xerox, and promoting its adoption in US industries. Since then, the application of QFD in the United States has been growing at a significant rate [9]. Dr ReVelle [10] has trained many engineers on QFD and Taguchi methods at Hughes Aircraft, Delco Electronics and other General Motors Divisions.

At the Ford Motor Company in 1989 alone, there were 80 to 90 QFD studies underway, and most of the 2600 engineers in the body and chassis group had taken training in QFD. Boeing, Hughes, Digital Equipment, Hewlett-Packard, AT&T, and ITT are also reported to be using QFD for a variety of products [3]. The use of QFD for computer software development is outlined by Chang [11] and Zultner [12].

5.5   THE QFD METHODOLOGY

QFD is a cross-disciplinary activity that can involve all pertinent organizational functions of a company, and is usually implemented by a mixed-discipline design team, consisting of members from marketing, design, manufacturing, logistics, maintenance and finance. The major work of conducting QFD is centred around filling a number of translation matrices to progress from the customer requirements to the detailed control of operations (Figure 5.2). The number of translation matrices is determined by the properties and complexity of the product, as well as by the level of detail required. However, the structure of all translation matrices is similar. The first translation matrix, which is often called the house of quality (HOQ), has the most general structure (Figure 5.3). In the HOQ there are six types of elements (Table 5.1). One example of a HOQ, Figure 5.4, shows the relationship between the customer requirements (ease of opening and closing door, and degree of isolation) to the specific design/engineering requirements (such as force to close door, and

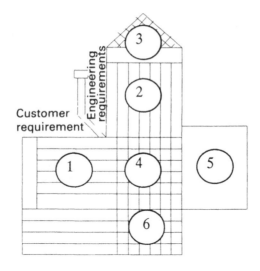

Fig. 5.3 House of quality – structure.

road noise reduction). Additional information is also given on the correlation between the engineering requirements (3 in Figure 5.3), the degree of competitiveness (5 in Figure 5.3), and on technical and cost assessments of the design/engineering requirements (6 in Figure 5.3). It should be noted that the structure of matrices vary from application to application. The QFD process commences by capturing the customer requirements of what the customers want in the product (1 in Figure 5.3) and how important each attribute is to the customer. In the car door example, Figure 5.4, what the customers wanted from a car door was 'ease of opening and closing' and 'isolation'. This information may not be detailed enough initially and may therefore have to be decomposed into a more detailed level. In this example, 'ease of opening and closing' can be decomposed into 'easy to close from outside', 'stays open on a hill' and so on.

Table 5.1 House of quality elements

Elements	Contents
1	Customer requirement list and relative prioritization (a What list)
2	Engineering characteristic list (a How list)
3	Engineering characteristic interrelation matrix or correlation matrix (roof)
4	Customer requirements – engineering characteristics interrelationships
5	Customer preference chart used to assess relative competitiveness
6	Technical and cost assessment used to allocate resource on elements

Having an accurate 'voice' from the right 'customer' is critical to the success of QFD. The 'voice of the customer' can be obtained through marketing surveys, marketing data, customer complaints, or by evaluating competitor's products. For some more intangible products, such as a computer system, the customers may not be obvious but may include, for example, both internal personnel, such as information system managers and data processors, and external customers, such as vendors. Since the company's objectives are to satisfy the customers needs, and to make a better product than that of its competitors, the design team must know where it stands relative to its competitors. The team can list this information on the right side of the HOQ (5 in Figure 5.3). Ideally, these evaluations are based on scientific surveys of customers. However, for some new products this information may not be readily available. The team documents the identified 'voice of the customer' (1 in Figure 5.3) and links this to the design requirements (2 in Figure 5.3), with relationships indicated in the matrix (4 in Figure 5.3). The objective measures at the bottom of the house are then added (6 in Figure 5.3). At the same time the team should pay attention to the roof of the house (3 in Figure 5.3), since one engineering change may affect other characteristics and this provides the basis for dialogue, assessment of trade-offs and decisions on compromise based on company priorities and competitive strategy.

At this stage the needed data for the entire first house of quality (HOQ) has been filled out. The design team can translate the design requirements into parts' characteristics through another HOQ or translation matrix. This process can continue to as detailed a level as desired and flow as far down-stream in the company product concept-to-customer sequence as appropriate for each product.

From the above, we can see that QFD is a structured decision-making framework that usually involves a cross-functional design team working together. Since QFD is being applied to the entire process of conceiving, developing, planning and producing a system, it is more than a quality tool [8, 9, 1]. It can be defined as a conceptual map of the elements, events and activities that are necessary throughout the development process to achieve customer satisfaction. It is a technique-orientated approach that uses surveys, reviews, analyses, relationship matrices and robust design, all centred on the theme of translating the voice of a customer into items that are measurable, actionable and potentially capable of improvement [9].

5.6   QFD AND SYSTEMS ENGINEERING

It is logical to think that the QFD approach is somewhat similar to the systems engineering (SE) approach. However, QFD is superior to SE for this application, in that QFD places an emphasis on understanding the

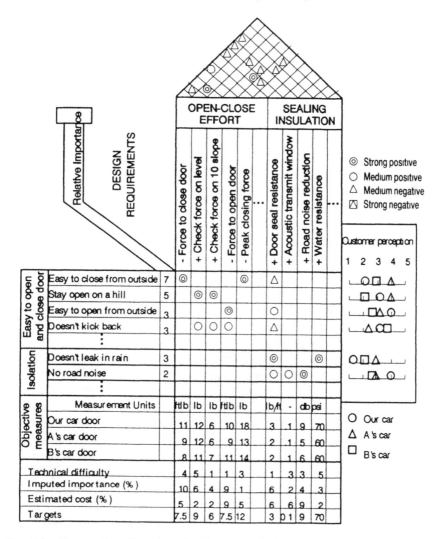

Fig. 5.4 House of quality example (Hauser and Clausing, 1988 [3]).

interrelationships among the requirements at the various levels of design [13, 9]. Thus QFD captures a great deal of design information which is lost in the SE approach. Other differences between QFD and SE are:

- QFD emphasizes not only the product but also the production process, while SE focuses efforts on the product.
- QFD reflects the 'voice of the customer', since the original customer requirements dictate the design activity. In contrast, the SE-generated 'requirements specifications' are considered to be the 'voice of the engineer', since the engineer generally takes responsibility for the requirements when using SE.

- QFD's interrelated matrices depict the design process and provide a means to reconcile the conflicting elements, whereas SE does not provide a mechanism for reconciling conflicts.

In both QFD and the SE approaches, requirements are thoroughly defined and the design follows a hierarchical decomposition methodology.

5.7   WHAT ARE THE MAIN ISSUES IN QFD?

Conducting QFD requires team work. The first and the most important issue for a successful application of QFD is the organization of the team. The team members should represent all functions related to the product, including marketing, design, manufacturing and planning. However, co-ordinating a multi-disciplinary team can be difficult. The level of involvement of the various team members is significantly different in the different translation stages, so that, for example, there is little involvement of the manufacturing expert in generating customer requirements. Therefore, maintaining team efficiency can be a problem. Another problem associated with the use of teams in QFD is the difficulty of maintaining communication amongst team members, with the semantics of one area having little overlap with the semantics of another.

The second major issue in applying QFD is the trade-offs needed between conflicting objectives. This is associated with the issue of team work, since team members often need to compromise to achieve a satisfactory consensus-based solution. Achieving such a goal-orientated team may need substantial education efforts in team skills and problem solving techniques, which many companies have adopted already as part of initiatives to implement concurrent engineering.

Another issue in applying QFD is the inadequacy of the support tools. The product development process is frequently so detailed and complicated that no one individual can comprehend it all, and the implementation of QFD can falter through the lack of suitable tools to guide the team through the maze of information. According to a fairly comprehensive survey of computer techniques in concurrent engineering [14] there is no work to date which adequately applies computer technology to QFD. A system called QFD/CAPTURE, developed by the International TechneGroup Incorporated, makes an attempt to assist in the documentation of QFD. However, the system does not entirely capture the logical structure of QFD. Another system called QFD Designer has been recently developed by American Supplier Institute. QFD Designer allows a user to build QFD charts, enter information on them, and navigate on different charts and on different levels. Overall, QFD is currently completely manual or is aided, via a static documentation process, by a computer system. This can greatly restrict the potential of QFD in that defects in design parameters, from CAD systems or from databases, cannot be automatically detected nor consistency automatically checked.

There is therefore a need for a comprehensive QFD computer system that can automatically check consistency and enable associativity to CAD systems and databases in a similar manner to the parametric infrastructure provided in the ProEngineer solids–CAD modelling environment.

REFERENCES

1. Akao, Yoji (ed.) (1990) *Quality Function Deployment – Integrating Customer Requirements into Product Design,* Productivity Press, Massachussets.
2. Bossert, J.L. (1991) *Quality Function Deployment – a practitioner's approach,* ASQC Press/Marcel Dekker publication, Milwauke, WI, USA.
3. Hauser, J.R. and Clausing, D (1988) The House of Quality. *Harvard Business Review,* **May–June,** 63–73.
4. Sullivan, L.P. (1986) Quality Function Deployment. *Quality Progress,* **June,** 35–50.
5. Langdon, R. (1990) Quality Function Deployment: New weapon in the zero-defects war. *Automotive Engineer,* **February–March,** 56.
6. Dale, M. and Best, C. (1988) Quality techniques in action, AE series on Engineering for Quality. *Automotive Engineer,* **Aug–Sep,** 44–8.
7. Eureka, W. and Ryan, N. (1988) *The Customer-Driven Company,* ASI Press, Dearborn, MI, USA.
8. Clausing, D. (1989) *Concurrent Engineering· Keynote Speech and Notes.* Concurrent Product and Process Design Symposium, ASME Winter Annual Meeting.
9. Schubert, M.A. (1989) *Quality Function Deployment – a Comprehensive Tool for Planning and Development.* Proceedings of the IEEE 1989 National Aerospace and Electronics Conference, V. **4,** 1498–503.
10. ReVelle, J.B. (1990) *The New Quality Technology – an introduction to QFD and Taguchi methods,* Hughes Aircraft Co., B138, Los Angeles.
11. Chang, Chia-hao (1989) Quality Function Deployment (QFD) Process in an Integrated Quality Information System. *Computers and Industrial Engineering,* **17,** (1–4), 311–16.
12. Zultner, R.E. (1989) Software Quality [Function] Deployment. *ASQC Quality Congress Transactions,* 558–63.
13. Hill, R.R. (1990) *Enhancing Concurrent Engineering Using Quality Function Deployment Based Tools.* Concurrent Engineering Symposium, 701–20.
14. Ishii, Kosuke (1990) *The Role of Computers in Simultaneous Engineering.* Proceedings of the 1990 ASME International Computers in Engineering Conference, 217–24.

CHAPTER 6

Design for manufacture

C. S. Syan and K. G. Swift

6.1 INTRODUCTION

Many industrial manufacturing problems and inefficiencies can be traced back to the design process. Substantial reductions in manufacturing costs can result from revisions at the design stage and such measures can crucially affect the success of a product. Often the benefits of a redesign are realized too late.

Through the creative work of the design engineer, the market need for a product is translated into drawings and a bill of materials. The engineer has to consider many conflicting and complex issues in the design activity. The product must function in the most efficient and economical manner within the applying constraints. The major constraint is cost, although other factors such as safety, pollution and legal requirements often have to be considered. What sets the successful designer apart is the ability to design an efficient product which can be economically produced.

All engineers know that design governs product performance, but rarely appreciate the impact it has on manufacture. Consideration of manufacturing problems at the design stage is the major way of reducing manufacturing costs and increasing productivity. This is clearly illustrated by the substantial reductions in manufacturing costs resulting from revisions at the design stage through value analysis or similar techniques. Value analysis consultants can safely guarantee to reduce the cost of any product by 15%, with savings of 30% being typical [1]. Too often the benefits of such design revisions are realized too late. If the product is already in production, the issue is complicated by factors such as the investment committed to tooling and equipment and the remaining product life expectation, whereas an early identification of improved design for manufacture can maximize the benefits.

There is a realization in industry that many manufacturing problems stem from product designs that are inherently difficult and expensive to manufacture and assemble. We know that in the production of

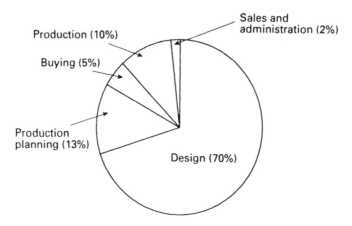

Fig. 6.1 Proportion of product costs determined by various departments.

engineering products, the activities of the design department largely predetermine how a product is made and what it will cost to produce [1] (see Figures 6.1 and 6.2). The product designer is normally limited only by a few critical dimensions and has much scope in the design of individual components and assemblies.

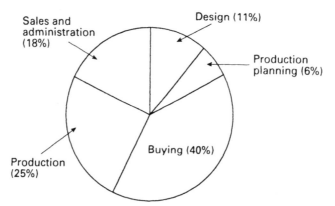

Fig. 6.2 Proportion of operating costs incurred in these departments.

The design process sets out to create products which fulfil a number of primary criteria, including the following.

1. Satisfaction of the functional and aesthetic requirements.
2. Optimum quality and reliability.
3. Compliance with company/legal policies on standardization and variety reduction.
4. Minimum part manufacture cost.
5. Minimum assembly cost.

Table 6.1   Design for manufacture – satisfaction of function at minimum cost

Create product design concepts to satisfy functional needs
Material selection: bulk, surface treatments or coatings
Design for assembly: assembly method, ease of handling and construction
Design for part manufacture: process selection, design for processing, tooling
   design
Quality and reliability
Standardization and variety reduction

Within criteria 1–4 the designer must consider factors relating to materials and surface treatment/coating selection (amongst many others) (Table 6.1), whilst 4. and 5. require specialized knowledge about the design features necessary for specific manufacturing and assembly processes.

6.2.   THE ROLE OF DFM IN CE

Design for manufacture (DFM) represents a new awareness of the importance of design as the first manufacturing step. It recognizes that a company cannot meet quality and cost objectives with isolated design and manufacturing engineering operations. To be competitive in today's marketplace requires a single engineering effort from concept to production.

The DFM approach embodies certain underlying imperatives that help maintain communication between all components of the manufacturing system and permit flexibility to adapt and to modify the design during each stage of the product's realization. Chief amongst these is the team or simultaneous engineering approach, in which all relevant components of the manufacturing system, including outside suppliers, are made active participants in the design effort from the start. The team approach helps ensure that total product knowledge is as complete as possible at the time each design decision is made. Other imperatives include a general attitude that resists making irreversible design decisions before they absolutely must be made and a commitment to the continuous optimization of product and process. The objectives of the design for manufacture approach are to identify product concepts that are inherently easy to manufacture, to focus on component design for ease of manufacture and assembly, and to integrate manufacturing process design and product design to ensure the best matching of needs and requirements.

Meeting these objectives requires the integration of an immense amount of diverse and complex information. This information includes not only considerations of product form, function, and fabrication, but also the organizational and administrative procedures that underlie the

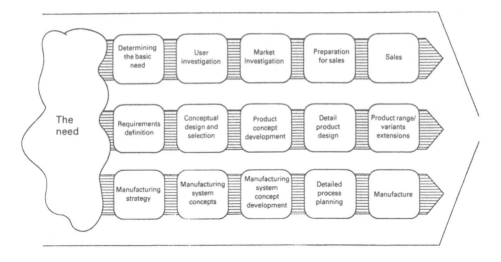

Fig. 6.3   A model of integrated product introduction.

design process and the human psychology and cognitive processes that make it possible.

Because of the complexity of the issues involved, it is convenient to divide the subject of DFM into two considerations:

1. Organizational and management issues
   These issues encompass project management and teamwork aspects. They must also include due consideration simultaneously to marketing, design, etc. Figure 6.3 (based on [1]) shows how by using the DFM approach or process, a product can be effectively designed for manufacture. The objective being to produce a design which is right first time.
2. Methodologies and tools
   These methodologies and tools can be used to help support DFM goals and to help ensure that the physical design meets the DFM objectives.

6.3   DFM METHODS

These can take many forms, ranging from simple rules of thumb through evaluation procedures to generative approaches to achieve manufacturing- and assembly-orientated designs. Table 6.2 includes some common approaches and indicates their main characteristics. As mentioned previously the use of design methods, or tools and techniques as they are often called, should help to guide the design team in their decision making relating to product function, manufacture and operational support.

Table 6.2 Some of the important methods influencing product costs in DFM

Methods	Characteristics
Value analysis/Engineering	Added value and cost reduction through function definition and function – cost analysis and teamwork
DFM guidelines (including design of components for manufacturability)	All encompassing rules for best practice for economic manufacture. Typically knowledge-intensive activity capable of methodical application for selecting appropriate manufacturing technology and design for ease of processing
Design for assembly	Rules specific to assembly, and other techniques including knowledge-intensive techniques which can be applied methodically for assembly rationalization, and design of components for ease of handling and assembly
Creative design methods	Divergent, non-patterned approach to problem solving, which can sometimes result in elegant solutions to previously intractable problems
Product family themes	Formation of component families by design or production-oriented coding systems or production flow consideration
Design axioms	Fundamental principles of good design
Poka Yoke/Taguchi design methods	Methods for design foolproofing for robust design and statistical design of experiments

A problem in understanding the role of the various techniques given in Table 6.2 is that while the techniques are different in character, there is much overlap and the results obtained can be very similar. For example, results widely quoted for DFMA [2] look much like those attributed to value engineering work [3].

Perhaps the key ingredients for the success of any DFM approach are firstly, and most importantly, a systematic application process, and secondly, the provision of useful knowledge on the best practice for the achievement of the required goals. A brief description of the various methods and a review of how they are used in the product introduction process follow.

6.3.1 Value analysis/engineering

Value analysis is a well-established technique which involves defining the functions performed by a product, its assemblies and components, and establishing the costs associated with these functions. In this way poor value and high cost functions can be identified. Alternative designs are then postulated and evaluated. The technique is broad in perspective, for it can contribute through brainstorming to creative thinking, and can also result in standardization of components [3]. Therefore, value analysis, or value engineering as it is called when applied to products before they reach production, avoids costly errors which are difficult to correct once a design is being produced and offers the benefit of a standardized solution virtually throughout the design process. For value analysis/engineering to succeed its application needs to be well organized and the team members must have expertise in manufacturing. Depending on the size and organizational philosophy of the company, these functions may be stand-alone departments, or they may be included as part of either the product design or manufacturing engineering team. Typically, therefore, value engineering is held responsible to a manager associated with design engineering.

6.3.2   Design for manufacture guidelines

Design for manufacture guidelines are basically concerned with optimizing the manufacturing system with respect to cost, quality and productivity. Specifically, they explain how the product design interacts with other components of manufacturing systems. This information is intended for use by designers and product engineers early in the design process. DFM guidelines are systematic and codified statements of good design practice that have been empirically derived from years of design and manufacturing experience. Typically, the guidelines are stated as directives that act to both stimulate creativity and show the way to good design for manufacture. If correctly followed, they should result in a product that is inherently easier to manufacture [4]. Various forms of the design guidelines have been stated by different authors, a sampling of the most commonly used are listed as follows:

1. Design for a minimum number of parts.
2. Develop a modular design.
3. Minimize part variations.
4. Design parts to be multi-functional.
5. Design parts for multi-use.
6. Design parts for ease of fabrication.
7. Avoid separate fasteners.
8. Minimize assembly directions; design for top-down assembly.
9. Maximize compliance; design for ease of assembly.
10. Minimize handling; design for handling and presentation.
11. Evaluate assembly methods.
12. Eliminate or simplify adjustments.
13. Avoid flexible components.

DFM guidelines show the way, but do not replace the talent, innovation and experience of the product development team. They must also be applied in a manner that maintains and, if possible, enhances product performance and marketing goals. Design guidelines should be thought of as 'optimal suggestions', which, if successfully followed, will result in a high-quality, low-cost and manufacture-friendly design. If a product performance or marketing requirement prevents full compliance with a particular guideline, then the next best alternative should be selected.

6.3.3   Design for assembly

DFA systems do more than analyse the ease of assembly and handling. They are based largely on industrial engineering time study methods. The DFA method developed by Hitachi, Boothroyd and Dewhurst, and Lucas seeks to minimize the cost of assembly within constraints imposed by other design requirements. This is done by first reducing the number of parts and then ensuring that the remaining parts are easy to assemble. Essentially, the method is a systematic, step-by-step implementation of the DFM guideline numbers 1., 7., 8., 9., and 10. Because of the level of interest in the DFA field, it is the subject of Chapter 7 of this book.

6.3.4   Creative design methods

Opportunities for improved competitiveness occurs at two different levels during product introduction. Firstly, at the conceptual design stage, where alternative working principles are generated to satisfy the product design specification. Secondly, during concept development and detailing, where within the framework of the design concept in question components are defined in detail and assembly and materials are specified. Research has indicated that within a working principle much variation can exist in design solutions, and many openings for part count reductions and design for processing have been reported [5]. Great scope can also lie in the generation of original design concepts, based on research and scientific knowledge, which are inherently economic to produce. Such original work can completely change the competitive position of a business.

BS 7000, *Guide to Managing Product Design* [6] was conceived in 1985 and became the British Standard in 1989. This standard identifies five basic pre-requisites for effective management: sincere commitment to product design, motivation of all those involved, provision of clear objectives, provision of adequate people and equipment resources and provision of an organizational system. This provides a very general guide to the difficult problem of successful product design and its management.

Specification of products has long been identified as a weakness in the European and US industries. Indication of the format of the product

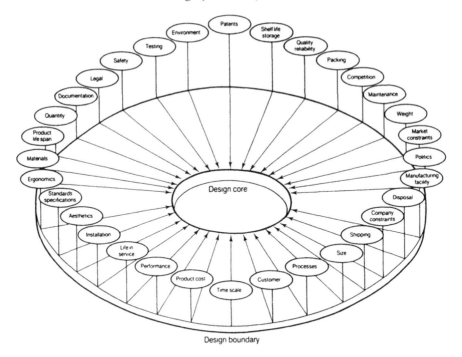

Fig. 6.4 Elements of the PDS [7].

design specification (PDS) is given by Pugh [7]. There is a recent British Standard aimed at addressing this issue, BS 7373, *Guide to the Preparation of Specifications* [8]. This standard provides extremely useful guidance for preparing a PDS in a very practical way. The success of products depends upon the comprehensiveness of the PDS. Pugh [7] recognized this fact and identified 30 elements (or trigger areas) which must be considered prior to the commencement of, or as part of, the initial design process. Figure 6.4 shows the basic elements of a PDS first enunciated in 1974. For further information, the reader is referred to *A Guide to Design for Production* [9] by The Institution of Production Engineers (now the Institution of Manufacturing Engineers, part of the Institution of Electrical Engineers).

6.3.5 Product family themes

Group technology (GT) is an approach to design and manufacture that seeks to reduce manufacturing system information content by identifying and exploiting the similarity of parts based on their geometrical shape and/or similarities in their production process. GT is implemented by utilizing classification and coding systems to identify and understand part similarities and to establish parameters for action.

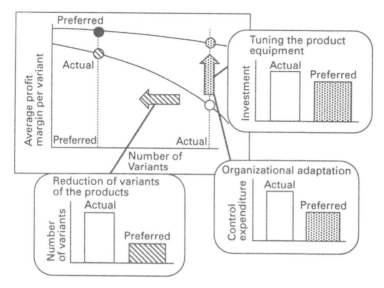

Fig. 6.5 Different ways of controlling the variety of variants.

As a DFM tool, group technology can be used in a variety of ways to produce significant design efficiency, product performance and quality improvements. One of the most rapidly effective of these ways is the use of GT to help facilitate significant reductions in design time and effort. In using a GT system, the design engineer needs only to identify the code that describes the desired part. A search of the GT database reveals whether a similar part already exists. If it does, and this is most often the case, then the designer can simply modify the existing design to design the new part. In essence, GT enables the designer to literally start the design process with a nearly complete design.

Group technology can also be effectively used to help control part proliferation and eliminate redundant part designs by facilitating standardization and rationalization approaches. If not controlled, part proliferation can easily reach epidemic proportions, especially in large companies that manufacture many different products and product models. By noting similarities between parts, it is often possible to create standardized parts that can be used interchangeably in a variety of applications and products.

Notable work in this area can be found in reference [10], which describes a variant mode and effects analysis (VMEA) for reducing the number of product variants. Figure 6.5 shows the various ways of controlling variants and Figures 6.6 and 6.7 illustrate the results of the VMEA analysis on the Audi 100 car pedal base. The original design had 67 parts and 16 variants, which were reduced to 47 parts and 6 variants. The overall savings produced from this exercise suggest that on average it is possible to save up to 16% of the total product costs.

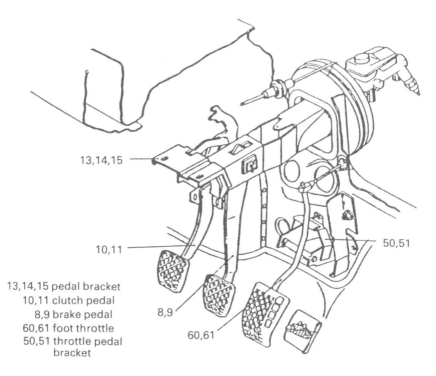

13,14,15

10,11

50,51

13,14,15 pedal bracket
10,11 clutch pedal
8,9 brake pedal
60,61 foot throttle
50,51 throttle pedal
     bracket

8,9

60,61

Fig. 6.6   Foot controls for the Audi 100.

6.3.6   Design axioms

An axiomatic approach to design is based on the belief that fundamental principles or axioms of good design exist and that use of the axioms to guide and evaluate design decisions leads to good design. By definition, an axiom must be applicable to the full range of design decisions and to all stages, phases and levels of the design process. Design axioms cannot be proven, but rather must be accepted as general truths because no violation or counter example has ever been observed.

A study of many successful designs by Suh, Bell and Gossard in 1977 led them to propose a set of hypothetical axioms for design and manufacturing [11]. Analysis and refinement of the initial axioms has shown that good design embodies two basic concepts. The first is that each functional requirement of a product should be satisfied independently by some aspect, feature or component within the design. The second is that good designs maximize simplicity; in other words, they provide the required functions with minimal complexity.

Use of design axioms in design is a two-step process. The first is to identify the functional requirements (FRs) and constraints. Each FR should be specified such that the FRs are neither redundant nor

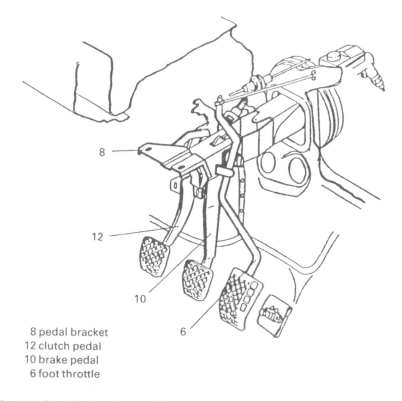

8 pedal bracket
12 clutch pedal
10 brake pedal
6 foot throttle

Fig. 6.7 Standardized car pedal system for the Audi 100

inconsistent It is also useful in this step to order the FRs in a hierarchical structure, starting with the primary FR and proceeding to the FR of least importance Once the functional requirements and constraints are specified for a given product or design problem, the second step is to proceed with the design, applying the axioms to each design decision, each decision should be guided by the axioms and must not violate them

Application of the design axioms to the analysis and design of products and manufacturing systems is not always easy or straightforward Because the axioms are quite abstract, their use requires considerable practice as well as extensive on-the-job design, manufacturing experience and judgment

Although several axioms were originally prepared, these have been reduced to the fundamental axioms and corollaries as shown in Table 6 3

6.3.7 Poka Yoke/Taguchi design methods

The provision of effective quality assurance to prevent defective products being produced is essential if economic and quality manufacture is to be

Table 6.3   The design axioms and seven main corollaries

Axioms	
Axiom 1	In good design the independence of functional requirements is maintained
Axiom 2	Among the designs that satisfy Axiom 1, the best design is the one that has the minimum information content
Some of the important corollaries	
Corollary 1	Decouple or separate parts or aspects of a solution if functional requirements are coupled or become coupled in the design of products and processes
Corollary 2	Integrate functional requirements into a single physical part or solution if they can be independently satisfied in the proposed solution
Corollary 3	Minimize the number of functional requirements and constraints
Corollary 4	Use standardized or interchangeable parts whenever possible
Corollary 5	Make use of symmetry to reduce the information content
Corollary 6	Conserve materials and energy
Corollary 7	A part should be a continuum if energy conduction is important

realized It is essential to ensure, for example, that a design cannot be assembled incorrectly [12] Poka Yoke provides techniques to ensure consideration of incorrect assembly early in product design process This is of course the rationale of foolproofing [13]

Taguchi design methods [14] address the problems associated with determining robust design by using statistical design of experiment theory Robust design implies a product is designed to perform its intended function no matter what the circumstances are In particular, the Taguchi method seeks to identify a robust combination of design parameter values by conducting a series of factorial experiments and/or using other statistical methods Termed parameter design by Taguchi, this step establishes the mid-values for robust regions of the design factors that influence system output

The next step, called tolerance (allowance) design, determines the tolerances of allowable range of variation for each factor The mid-values and varying ranges of these factors and conditions are considered as noise factors and are arranged in orthogonal tables to determine the magnitude of their influences on the final output characteristics of the system A narrower allowance will be given to noise factors, imparting a large influence on the output

In establishing the tolerance or allowance range for a particular parameter, Taguchi uses a unique concept defined as a loss function In this approach, loss is expressed as a cost to either society (the customer) or the company that is produced by deviation of the parameter value from design intent. Because any deviation from design intent produces a loss, allowance or permissible deviation should be determined based on the magnitude of the cost associated with this loss. The concept of loss and other Taguchi concepts provide valuable insight into quality and the role design plays in determining the quality of a product or system.

6.4   COMPUTER-BASED APPROACHES TO DFM

A major perceived barrier to DFM application in companies is the additional time requirement. Design and manufacturing engineers are typically operating under very tight schedules and are therefore reluctant to spend time learning and using DFM approaches. Computer-aided DFM helps simplify the effort and shortens the time required to implement DFM on a daily basis. Computer-aided DFM also enables the design team to consider a multitude of product/process alternatives easily and quickly. 'What-if' optimization allows each alternative to be refined and fine tuned. Together, these capabilities greatly increase the probability of identifying the most desirable solutions during the early stages of design. When properly implemented an applied, computer-aided DFM has the potential to vastly improve the quality of early product/process decisions and thereby enhance the design team's ability to design for effective quality, cost and delivery. Computer-aided DFM can also foster team building and a team approach.

The benefits of computer-based DFM as compared with manual application include:

1. Paperless operation.
2. Less tedious and time saving.
3. Consistency of results and their presentation.
4. More knowledge may be used without overpowering the user.
5. Possibility of automatic evaluation and redesign.

A variety of proprietary computer-aided DFM software packages are currently available for parts of the DFM field. Boothroyd Dewhurst Incorporated (BDI) sell modules for manufacturability analysis to supplement their design for their assembly (DFA) package in casting, fabrication and electronics manufacturing [15]. Lucas Engineering Systems have a graphics-based DFA system incorporating analysis for component manufacturing, process selection and casting. In addition, considerable effort is being directed towards the development of new computer-based and/or computer-aided DFM methodologies in research and development institutions. However, DFM is a very wide and complex field and the computer-based systems address small parts of the total problem, and hence are at the present of limited practical use.

6 5 DISCUSSION AND CONCLUDING REMARKS

As stated above, DFM is a wide and non-trivial subject A large number of methods and non-standard terminology has been developed Hence, a commonly agreed approach, and application methods for DFM implementation, are not well defined

One point is clear – the need for training of engineers in academia and industry in this important area Users should be aware of the impact that design has on manufacturing and introduced to the theory and practice of techniques discussed in this chapter The use of case history material is essential for developing the practical proficiency of the users DFM should be given a high priority in companies and an appropriate emphasis in the training portfolios

The research and developments in DFM promise easier paths for implementation and will encourage wider adoption of principles The work in integration of procedures to enable the evaluation of component design for manufacturing cost can provide a holistic means of identifying weaknesses in product design solutions

Work on the computer-based DFM systems is being augmented by the increasing use of knowledge-based expert systems approaches These offer natural methods of representing expert knowledge and a means of emulating expert problem solving, as there are no easy algorithmic solutions to the DFM problems They need to be integrated with other tools and systems used in the concurrent engineering (CE) approach to realize fuller benefits to businesses

REFERENCES

1 Andreason, M , Myrup and Hein, L (1987) *Integrated Product Development*, IFS Publications Ltd , Springer, Berlin
2 Boothroyd, G and Dewhurst, P (1991) Product Design for Manufacture and Assembly, in *Design for Manufacture* (eds Corbett, J , Dooner, M , Meleka, J and Pym, C), Addison-Wesley Publishers, Wokingham, UK
3 Raven, A D (1971) *Profit Improvement by Value Analysis Value Engineering and Purchase Price Analysis*, Cassell Publishing, London, UK
4 Stoll, H W (1988) Design for Manufacture *Manufacturing Engineering*, January, 67–73
5 Tuttle, B L (1991) *Design for Function A Cornerstone for DFMA* International Forum on Product Design for Manufacture and Assembly, June 1991, Newport, Rhode Island, USA
6 BSI (1989) *BS 7000 – Guide to Managing Product Design*, BSI, Milton Keynes
7 Pugh, S (1991) *Total Design*, Addison-Wesley Publishers, Wokingham, UK
8 BSI (1991) *BS 7373 – Guide to The Preparation of Specifications*, BSI, Milton Keynes
9 The Institution of Production Engineers (1984) *A Guide to Design for Production*, IProdE Publication, UK
10 Schuh, G and Becker, T (1989) *Variant Mode Effect Analysis A New Approach*

for Reducing the Number of Product Variants, Fourth Boothroyd-Dewhurst Inc. Conference, Rhode Island 1989, USA.

11. Suh, N.P., Bell, A.C. and Gossard, D.C. (1978) On an Axiomatic Approach to Manufacturing and Manufacturing Systems. *ASME Journal of Engineering for Industry*, **100**(2).

12. Parnaby, J. (1989) *Lucas Mini-Guide 3, Total Quality*, Lucas Engineering and Systems, Birmingham, UK.

13. Schneider, W. (1990) *Generation of Assembly Oriented Design*. Proceedings of Institution of Mechanical Engineering Seminar on Design for Manufacture and Quality within Integrated Product Development, April 1990, London, UK.

14. Taguchi, G. (1986) *Introduction to Quality Engineering: Designing Quality Products and Processes*, Kruef International Publishers, New York.

15. BDI Incorporated, (1993) 138 Main Street, Wakefield, Rhode Island, 02879 USA.

CHAPTER 7

Design for assembly

C. S. Syan and K. G. Swift

7.1 AUTHORS' NOTE

It is acknowledged that assembly represents one of the major factors affecting product cost and quality. Design for assembly (DFA) is a key element in the creation of competitive products and reducing time to market. In this chapter approaches to DFA, including 'assembly evaluation methods' are outlined and their implementation in product introduction is described.

7.2 INTRODUCTION

Products may consist of thousands of components and involve many different technologies. The conventional design process is largely sequential, with production method design coming after the design is finalized. In the design phase all components are designed in detail and the materials, surfaces and tolerances are specified. This means that the process and assembly methods are to a large extent already determined by the design process. Experts may say that 75% to 85% of the cost of a product is committed during the design and planning activities [1, 2]. Therefore, consideration of assembly problems at the product design stage is the most effective way available for reducing assembly costs and increasing productivity.

In designing for assembly and functionality there is a need to use information of considerable breadth and complexity. Expertise is required in many fields, including manufacturing engineering. DFA is thus best carried out using a variety of expertise in a well managed team-based environment. It therefore makes sense to address DFA from organizational as well as technical viewpoints. One has to think about adequate organizational structures and procedures to support DFA (see Chapter 3) and about methodologies and tools which can aid and accelerate the DFA process.

7.3   AIMS OF DESIGN FOR ASSEMBLY

The objective of DFA is to identify product concepts which are inherently easy to assemble and to favour product and component designs that are easy to grip, feed, join and assemble by manual or automatic means. This objective is related to the overall design for manufacture (DFM) approach to economic production. DFA can be carried out throughout the product introduction process from conceptual design to component detailing. The main aims of DFA are to:

1. reduce the number of parts in an assembly.
2. optimize the 'assemblability' of the parts.
3. optimize the 'handlability' of parts and assemblies.
4. improve quality, increase efficiency and reduce assembly costs.

DFA may be carried out manually or with the support of computers. Computer-based DFA systems are available as conventional programs or as knowledge-based expert systems.

DFA has traditionally relied on general guidelines and examples for the designer. More recently, works on DFA have concentrated on the evaluation of 'assemblability' in order to facilitate design improvements. Systems have been developed that enable designers and production engineers to measure the ease or difficulty with which components can be handled and assembled. Most of them operate by guiding the user through a systematic analysis in which the problems associated with components and processes are quantified. Evaluation results can be used to compare alternative design solutions or competitors' products.

'Assemblability evaluation' is generally carried out on completed product designs, existing products or prototypes, but 'assemblability analysis' is also possible at the conceptual drawing stage. This is preferred to forcing the designer to reject and re-design. In this chapter, methods for assemblability evaluation are introduced and discussed. These methods provide a measure of the suitability of a design for assembly and automation.

7.4   DFA TOOLS AND TECHNIQUES

In this section different approaches to DFA will be introduced. The emphasis is on giving the reader an understanding of the approaches available and their strengths and weaknesses.

7.4.1   Design for assembly guidelines

DFA guidelines tend to be rules learned from experience and developed over time, largely from the industrial assembly automation practice.

Generally, these rules take the form of sample lists with no framework or systematic guidance for their application. The guidelines are aimed at alerting the design engineer to points that should be considered in the design process. A checklist for DFA based on Corbett [3], includes the following:

1. Minimize:
   (a) parts and fixings
   (b) design variants
   (c) assembly movements
   (d) assembly directions.
2. Provide:
   (a) suitable lead-in chamfers (radii on corners)
   (b) natural alignment
   (c) easy access for locating surfaces
   (d) symmetrical parts, or exaggerate asymmetry
   (e) simple handling and transportation.
3. Avoid:
   (a) visual obstructions
   (b) simultaneous fitting operations
   (c) parts which will tangle or nest
   (d) adjustments which affect prior adjustments
   (e) the possibility of assembly errors.

Other workers [1, 4, 5] have quoted lists of guidelines. The ten most frequently quoted are:

1. **Minimize the number of parts.** Eliminate or combine them whenever possible. Non-existent parts do not have to be manufactured, handled, purchased, stored, inspected or serviced.
2. **Minimize assembly surfaces.** Simplify the design so that fewer surfaces need processing, and all processes on one surface are completed before moving to the next one.
3. **Design for top-down assembly.** This takes advantage of gravity to assist in assembly. Usually, tooling is less expensive, and fewer clamps and fixtures are needed.
4. **Improve assembly access.** Design for easy access, unobstructed vision and adequate clearance for standard tooling.
5. **Maximize part compliance.** Because of variability in manufacturing and tolerance, stack design with adequate leads, guide surfaces and specifications for mating parts to reduce misalignment.
6. **Maximize part symmetry.** Symmetrical parts are easier to orientate and handle. If symmetry is not possible, include obvious asymmetry or alignment features.
7. **Optimize part handling.** Design rigid, rather than flexible, parts

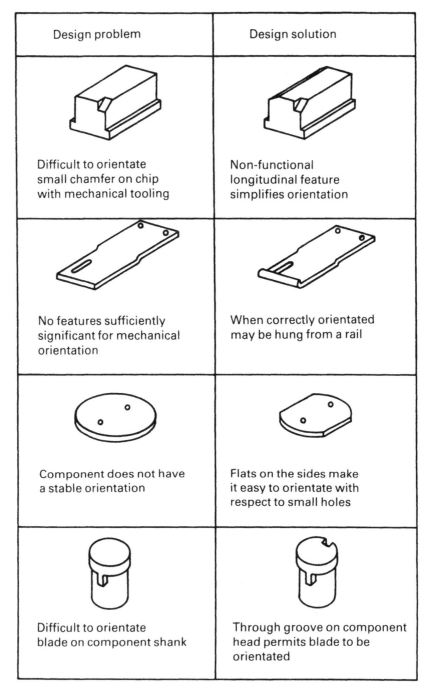

Design problem	Design solution
Difficult to orientate small chamfer on chip with mechanical tooling	Non-functional longitudinal feature simplifies orientation
No features sufficiently significant for mechanical orientation	When correctly orientated may be hung from a rail
Component does not have a stable orientation	Flats on the sides make it easy to orientate with respect to small holes
Difficult to orientate blade on component shank	Through groove on component head permits blade to be orientated

Fig. 7.1   Provision of asymmetrical features to assist in orientation.

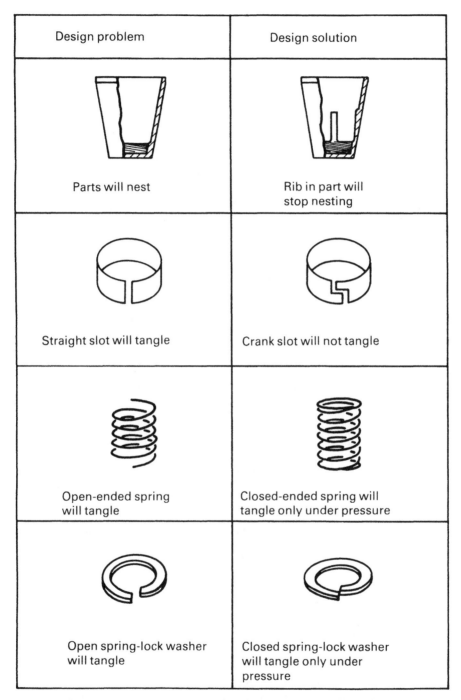

Design problem	Design solution
Parts will nest	Rib in part will stop nesting
Straight slot will tangle	Crank slot will not tangle
Open-ended spring will tangle	Closed-ended spring will tangle only under pressure
Open spring-lock washer will tangle	Closed spring-lock washer will tangle only under pressure

Fig. 7.2 Examples of re-design to prevent nesting or tangling of components.

Product design principle	Original design	Enhanced design
Design rationalization (13 components replaced by just 2) (a)	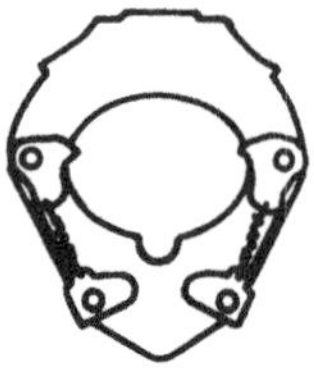	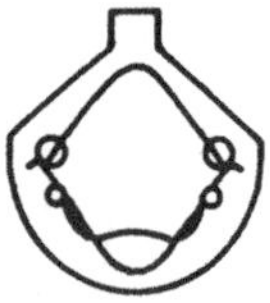
Provide guide on surfaces to aid component placing (b)	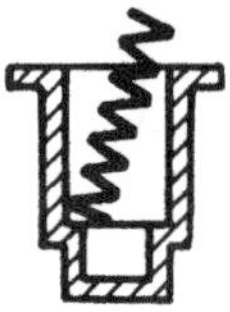	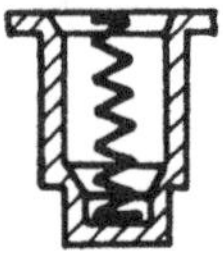
Location of base component in the machine workcarrier (c)		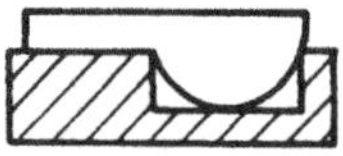
Design to allow assembly construction by simple linear motion (d)	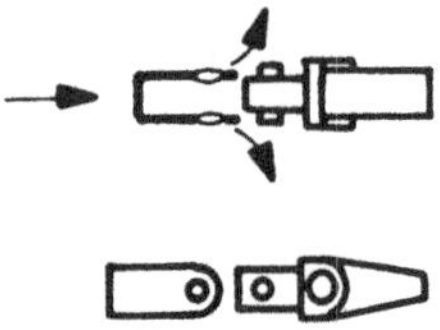	
Standardization of design (e)	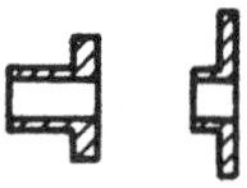	

Fig. 7.3 Illustrative examples of some of the principles of product design for automatic assembly.

where possible to ease handling. Provide adequate surfaces for gripping. Provide barriers to tangling, nesting or interlocking.

8. **Avoid separate fasteners.** Incorporate fastening into components, such as snap fits.

9. **Provide parts with integral self-locking features.** Provide tabs,

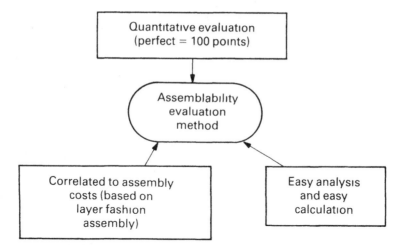

Fig. 7.4 General Hitachi assemblability evaluation method

indentations, or projections on mating parts to identify them and
their orientation through to final assembly

10 **Drive toward modular design and standardization.** Use standard
modules for common functional requirements, and standard interfaces for easy interchangeability of modules This allows more
options, faster updates of designs, easier testing and service
Standardize the fasteners that are used to reduce variation and
ensure availability

Some illustrative examples of design problems and their solutions, for
manual and automatic assembly, are given in Figures 7 1, 7 2 and 7 3 [6]

7.4.2 The Hitachi assemblability evaluation method (AEM)

The Hitachi model makes use of assembly cost ratios to identify the weak
points of a design The general AEM notions are given in Figure 7 4, and a
diagram illustrating the main steps in the evaluation procedure can be
found in Figure 7 5

The analyst completes an AEM form from drawings or samples by
entering the part names and numbers in the same order as an appropriate
structure of assembly work The assembly processes are analysed using
AEM symbols and calculated evaluation indices Scores (P) are obtained
for each part, and the product assembly evaluation score (E) is determined The evaluation score may also be correlated to an assembly cost
ratio (K) This step is followed by the judgemental stage, where the
evaluation indices are compared with target values (see Figure 7 5)
Finally, design improvement follows, where necessary, using the sequence (1)–(5) illustrated in the lower part of Figure 7 5

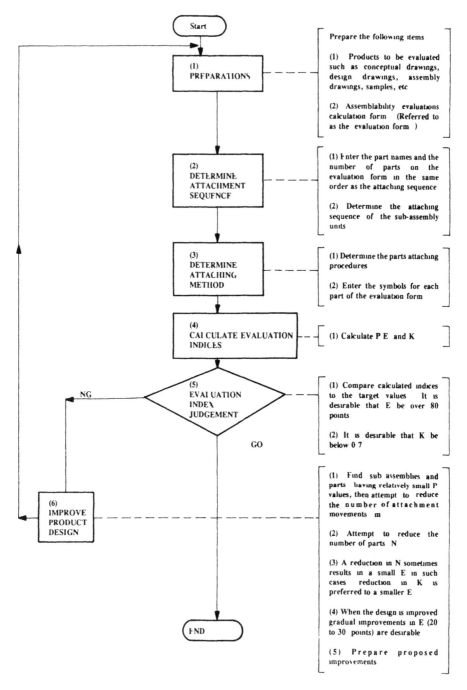

Fig. 7.5   The Hitachi assemblability evaluation procedure

Table 7.1 A conventional and a new distributor mechanism (photoelectric crank angle sensor built-in type)

Model	Current model (D4P88–03A)	Low height model (new) (D4P90–02)
Year of development	1988	1992
Size (volume)	100%	60%
Number of parts	60 (100%)	44 (73%)
Ease of assembly evaluation score	70	75
Important improvement	–	Abolishment of a leadwire and connectors assembly. Layer fashion structure. Reduction in number of parts.

The AEM does not distinguish between manual, robotic (flexible) and automatic (dedicated) assembly. Two reasons for this are given: the strong correlation between the degree of assembly difficulty involved in manual, robotic and automatic assembly, and the uncertainty involved in predicting the production mode in the early stages of product development. Production of the same design could well be moved from manual to some form of automated assembly, or vice versa.

Hitachi claims that the system is simple and easy to use, and that it saves tens of millions of dollars annually. Its application is mandatory within the company. Computer software versions are available which carry out the calculations and aid paperless analysis. For further information on this system, see Miyakawa and Ohashi [7]. An example application illustrating the benefits that can be achieved using this method are given in Table 7.1. The example shows a distributor mechanism assembly before and after analysis using the AEM technique. Many other Japanese companies use formal DFA methods and it is worth noting that Sony has a method very similar to Hitachi.

7.4.3   The Boothroyd Dewhurst DFA procedure

The design for assembly procedure marketed by Boothroyd Dewhurst Inc. [8], which is mainly aimed at mechanical assemblies, draws a sharp distinction between manual, robotic and high-speed automatic assembly. Indeed, there are separate analysis systems for each of these areas. The analysis section to be used is identified from a procedure for the selection of the appropriate assembly method. An idea of the most economical assembly method is found by considering factors such as production volume, number of parts to be assembled, number of product styles and payback period required.

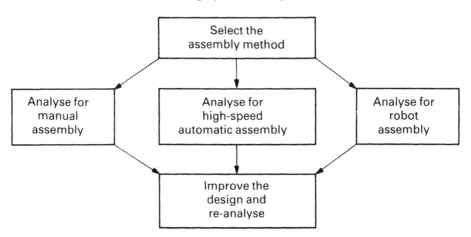

Fig. 7.6 Stages in BDI design for assembly analysis.

The Boothroyd Dewhurst DFA Handbook [4] suggests that the best way to achieve assembly cost reduction is to first reduce the number of components that must be assembled and then to ensure that the remaining components are easy to assemble. The basis for component reduction is that combining two components into one will eliminate at least one manual operation, or an entire section of an assembly machine. The analyses involved in this handbook are illustrated in Figure 7.6.

The next step is to apply the appropriate analysis system, i.e. manual, robotic or dedicated automatic assembly. In these procedures, classification systems, developed specifically for each of these technologies, are used to arrive at the data that is to assess the components in a design for ease of handling and insertion. Cost analysis for manual handling and insertion is based on estimating manual assembly costs using time data corresponding to particular component design classifications and operator wage rates.

In the analysis for automation, the classification systems for handling and insertion provide cost indices for component classes. These give an indication of the relative cost of the equipment required to automate the process, compared with the cost of equipment needed to process the most simple design. For the estimation of automated handling and insertion costs, it is necessary to know the cost of the equipment to process the most simple design. Having a relative cost index for a design, a basic cost for the automation equipment and an assembly production rate, some estimate of automated handling or insertion costs can be calculated.

The BDI analysis is carried out on a DFA form. Each of the three technologies has its own special form. A row of the form is completed for each component in that assembly. A blank automated assembly worksheet is illustrated in Figure 7.7. The completed worksheet provides a quantitative way of measuring the performance of a design in terms of its

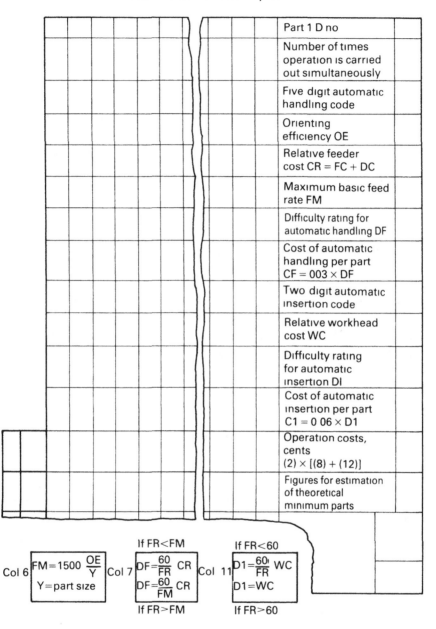

Fig. 7.7 Design for automatic assembly worksheet

assemblability, and can be used as a basis for comparing alternative designs. Table 7.2 shows the summary of the EPSON MX80 dot matrix printer which was marketed for use with PC and XT range of computers, and the summary of a different design printer, the IBM Proprinter. Analysis of these two devices using the BDI method for manual assembly clearly shows that for the IBM Proprinter assembly efficiency, time, cost

Table 7.2 BDI analysis of the EPSON MX80 and IBM proprinter products

EPSON MX80 final assembly		*Proprinter summary*	
Assembly efficiency (percent)	5	Assembly efficiency (percent)	51
Total assembly time (seconds)	552	Total assembly time (seconds)	170
Total labour cost (cents)	383	Total labour cost (cents)	118
Total number of operations	57	Total number of operations	32
Number of parts or sub-assemblies	49	Number of parts or sub-assemblies	32
Theoretical minimum number of parts or sub-assemblies	10	Theoretical minimum number of parts or sub-assemblies	29
Labour rate (dollars/hour)	25	Labour rate (dollars/hour)	25

and number of parts are all significantly lower than the MX80, illustrating the ability of the BDI method to evaluate different designs [9].

A computer-based toolkit version of the Boothroyd Dewhurst DFA evaluation is available with which the user can analyse the effects of a proposed design by editing the analysis. Assembly machine evaluation and design for automatic handling are also available as supplementary modules. Boothroyd Dewhurst Inc. produce a separate computer programme which extends the design for manual assembly evaluation to printed circuit board manufacture.

7.4.4 The Lucas DFA technique

The Lucas procedure arose out of the desire to have the best features of the commercially available packages incorporated within a user-friendly and flexible system, with the aim of application early in the design process. The structure and expertise used in this system are a result of knowledge engineering in organizations whose business is to manufacture and market assembly systems [10, 11]. The Lucas procedure uses common analysis models for manual and automatic systems, except in the case of component handling.

There are essentially three stages of analysis, each producing its own measure of performance, MOP (Figure 7.8).

1. Functional analysis is carried out according to the rules of value analysis, and components are divided up into those which have high functional priority and those groups which are of low functional importance. The analysis categorizes all components into A parts (demanded by the design specification) or B parts (required by that particular design solution) and sets a target of 60% for design efficiency (the number of A parts divided by A plus B, expressed as a percentage). The teams aim to exceed the 60% target value by the elimination of category B parts through re-design. This not only has a significant impact on direct costs but also reduces the indirect costs

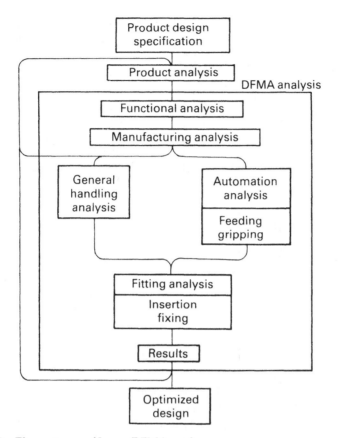

Fig. 7.8 Three stages of Lucas DFMA analysis.

incurred by every part from the moment of conception, such as the costs of documentation, process planning, procurement, quality assurance, production control, storage and distribution.

2. Handling analysis uses a knowledge base (paper or computer software) to assess the relative cost of handling every part, and sets a target for handling cost per A category part.

3. Likewise, fitting analysis uses another knowledge base to determine the cost of assembling every part and sets a target for assembly cost per A category part.

Armed with the analysis results and measures of performance, product re-design takes place to eliminate the high relative cost values and improve the MOPs. Re-analysis to validate the improvement is then the final step in the DFA procedure.

During fitting analysis the proposed assembly process is declared in the form of an assembly sequence flow chart. This declaration by the team at a very early stage in the design process actively promotes concurrent

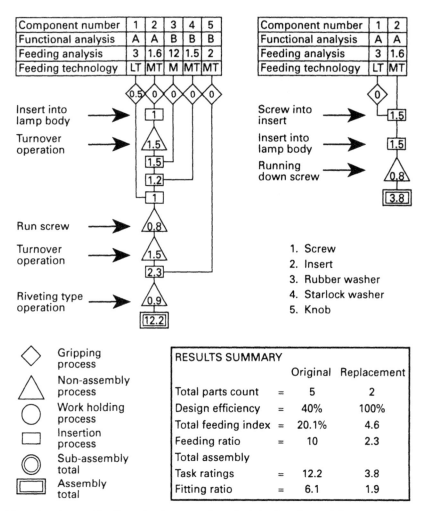

Component number	1	2	3	4	5
Functional analysis	A	A	B	B	B
Feeding analysis	3	1.6	12	1.5	2
Feeding technology	LT	MT	M	MT	MT

Component number	1	2
Functional analysis	A	A
Feeding analysis	3	1.6
Feeding technology	LT	MT

Insert into lamp body → 1

Turnover operation → 1.5

1.5

1.2

1

Run screw → 0.8

Turnover operation → 1.5

2.3

Riveting type operation → 0.9

12.2

Screw into insert → 1.5

Insert into lamp body → 1.5

Running down screw → 0.8

3.8

1. Screw
2. Insert
3. Rubber washer
4. Starlock washer
5. Knob

◇ Gripping process
△ Non-assembly process
◯ Work holding process
▭ Insertion process
◎ Sub-assembly total
▭ Assembly total

RESULTS SUMMARY		Original	Replacement
Total parts count	=	5	2
Design efficiency	=	40%	100%
Total feeding index	=	20.1%	4.6
Feeding ratio	=	10	2.3
Total assembly Task ratings	=	12.2	3.8
Fitting ratio	=	6.1	1.9

Fig. 7.9   Assembly flow chart comparison (original and new trim screws).

engineering and by its visibility encourages all team members to participate in decisions on product design and assembly.

The analysis results are displayed in the form of a combined table of results and assembly sequence flow chart. Figure 7.9 shows 'before' and 'after' results of a simplified illustrative example of an assembly that was reduced from five to two parts by re-design. The results box in the bottom right-hand corner compares the calculated measures of performance. This clearly demonstrates that the design represented by the 'after' flow chart is better than that represented by the 'before' flow chart. The reduction in the extent and complexity of the flow chart is also a good indicator, showing a preference for the re-designed product.

AIDS FOR ASSEMBLY-ORIENTED PRODUCT DESIGN / STEPS IN THE DESIGN PROCESS	Creative						Corrective					
1 PLANNING STAGE												
1.1 recognition of requirements					●							
1.2 definition of tasks	●				●		●	●	●			
1.3 create a requirements list											●	
1.4 release for rough design												
2. ROUGH DESIGN												
2.1 analysis of functions					●							
2.2 create functional structure	●										●	
2.3 create variations of functional structure	●									●	●	
2.4 determine solutions for each function	●	●	●		●	●						
2.5 select principles										●	●	
2.6 prepare solution variants										●		
2.7 work out different concepts	●	●	●	●		●						
2.8 evaluate and select the different concepts						●	●	●	●	●	●	●
3. DRAFTING STAGE												
1.3 drafting the main functional units	●	●	●	●								
3.2 drafting the remaining functional units	●	●	●	●								
3.3 select suitable parts of the draft							●	●	●			●
3.4 detailed design of the main and auxiliary functional units	●	●										
3.5 check and improve drafts				●							●	●
3.6 trace faults and problems												
3.7 analysis of cost recovery							●					
3.8 draft completion	●	●		●								●
3.9 decision or draft							●			●	●	●
4. FINAL DESIGN												
4.1 detailing												
4.2 working out specifications	●	●	●	●								
4.3 analysis of production data				●		●						
4.4 release for production							●					●

Fig. 7.10   Aids for assembly-oriented product design.

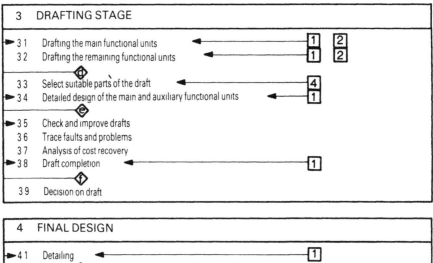

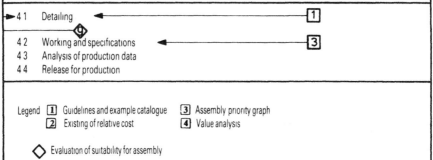

Fig. 7.11 Assembly-oriented design process

7.4.5 Assembly-oriented design process (generation techniques)

This is a structured approach to the application of the DFA guidelines [12, 13]. It is vital that the possibilities for influencing the suitability of assembly for a design be supported by the right aids at the right time. In Figure 7.10 the aids for assembly-oriented product design are listed in accordance with the different design stages. The necessary technical requirements of assembly are included at each stage during the design process for use on an iterative basis. This means that at the various stages of the process there will be the possibility of checking the suitability for assembly, facilitating continuous improvements. Figure 7.11 specifies the allocation of the aids detailed in Figure 7.10 and summarizes them for the assembly-orientated design process [14].

(a) Design rules for assembly-oriented product design

These rules represent a set of known and well-tried solutions for certain design tasks and can be conveniently summarized in catalogues. In the catalogue, the classification part is divided up into:

1. Measures for the product structure.
2. Measures for sub-assemblies.
3. Measures for individual components.
4. Measures for jointing techniques.

In the main part of this catalogue, the actual design rules are listed in detail, as shown by an extract in Figure 7.12. The access characteristics for this catalogue are:

- validity of the rules during the design process;
- importance of the rules, represented by the A, B and C importance factors attached to the rules.

Both characteristics are installed in a CAD environment so that the designer can call up the design rule catalogue through the CAD system's user interface and select the appropriate rules for his or her design problem. Optionally, the CAD system presents related DFA examples to the designer so that the design rules are visualized. The design rules catalogue contains about 120 rules [14] and is stored in the ORACLE database and Euclid CAD system.

(b) Assembly information system

The assembly information system provides the designer with necessary information in the early stages of the design process to avoid any crucial manufacturing and assembly errors. Technical and cost information

RULES FOR ASSEMBLY-ORIENTED PRODUCT DESIGN

2	Measures for sub-assemblies		
2.3	Design of the basic wall		
No:	Design rules	Importance	Design step
2.3.1	Mounting of sub-assemblies on the basic unit (preferably complete sub-assembly on the one basic unit)	A	11.7 111.1
2.3.2	Basic unit must have good positional stability	B	111.3
2.3.3	Basic unit must have a preferably large and plane surface	B	111.7 1111.1

Fig. 7.12 Extract from design rules for assembly-oriented product design.

about equipment in the field of automatic assembly already in use or available on the market is provided. For example, handling devices, grippers, automatic tools (e.g. automatic screwers, riveting tools), part feeding devices (e.g. bowl feeders, magazines) and sensors. Such support can speed up the design process and therefore save time and money.

(c) Evaluation of suitability of products for assembly

In order to evaluate the suitability of products for assembly, it is necessary to consider and evaluate all the factors of influence, i.e. the entire spectrum of assembly-oriented product design.

In the case of the assembly-oriented design process shown in Figure 7.9, the (a) and (b) evaluations were carried out with checklists which examine precisely those factors which are important during the various stages. As from evaluation (c), the suitability for assembly can be evaluated with an appropriate procedure The basic idea of this procedure is that the products must be designed to fulfil the requirements and functions listed on the specification sheet

Application of this methodology has shown [12] that precise analysis of the assembly process at the design stage is effective A method of assessment of the suitability for assembly of products or designs is a prerequisite for discovering and removing weak points This method of assessment has been developed with the aid of computers and implemented on PCs, so that the work stages 1–7 of the method can be rapidly executed, providing rapid application of DFA tools in design

7 5 CONCLUDING REMARKS

While design rules and principles represent a set of known and well-tried solutions for certain design tasks, their application has traditionally not been accompanied by a well-defined, systematic application procedure and quantifiable measures of performance These weaknesses have led to the development of the DFA evaluation methodologies These techniques facilitate quantification of the ease or difficulty of assembly and are useful in the assessment of design alternatives and complete products DFA does more than its name suggests, since part-count reduction, a key element of value analysis/engineering, has a major influence in the wider sphere of DFM

However, one of the major weaknesses of evaluation methods is that they all need detailed information which is only available towards the end of the product design development process Thus, in case of a poor evaluation result, effort has to be expended generating a better design, increasing costs and protracting time to market To avoid this, generation support methods for design for assembly, pioneered at IPA and partly realized as modules of the so-called 'design support system (DSS)', offer much promise in this area Some developments have also been made in the area of DFA/CAD integration ([15]) This area of work, although very important in DSS in DFA, has had limited success

REFERENCES

1 Andreason, M M , Kahler, T L and Swift, K (1988) *Design for Assembly* (2nd edn), IFS Publications, Springer-Verlag, Kempston UK
2 Sheldon, D F Perks, R (1990) Designing for Whole Life Cycle Costs *Journal of Engineering Design* **1** (2)
3 Corbett, J Dooner, M , Meleba, J , Pym, C (1991) *Design for Manufacture, Strategies Principles and Techniques*, Addison Wesley Publishers, Wokingham UK
4 Boothroyd, G and Dewhurst, P (1987) *Product Design for Assembly Handbook*, Boothroyd Dewhurst Inc , Wakefield, Rhode Island, USA

5 Owen, A E (1983) Designing for Automated Assembly *Engineering,* **June,** 436–41
6 Swift, K G (1986) A Computer-Based Methodology for Advising the Designer Regarding Assembly Automation, PhD thesis, University of Hull, UK
7 Miyakawa, S and Ohashi, T (1986) *The Hitachi Assemblability Evaluation Method (AEM)* International Conference on Product for Assembly, April 1986, Newport, RI, Pub Troy Conferences
8 BDI Incorporated, 138 Main Street, Wakefield, Rhode Island, 02879, USA
9 Dewhurst, P and Boothroyd, G , Design for Assembly in Action, *Assembly Engineering,* Jan, 1987, USA
10 Swift, K G and Miles, B (1985) *Design for Assembly, Vol 2,* Lucas Engineering and Systems Ltd , Solihull, Birmingham, UK
11 Miles, B L (1989) Design for Assembly – A Key Element for Design for Manufacture *Proceedings of the Institute of Mechanical Engineers,* **203** (D)
12 Warnecke, H K and Bassler, R (1988) Design for Assembly – Part of the Design Process *Annals of the CIRP,* **37**/1/1988
13 Schneider, W (1990) *Generation of Assembly Oriented Design* Proceedings of Institution of Mechanical Engineers' Seminar on *Design for Manufacture and Quality within Integrated Product Development,* UK
14 Bassler, R (1988) *Integrartion der Montagegerechten Produkgestaltung in der Konnstruktionsprozes,* Stuttgart Universssitat, Facultat Fertigungstechnik, Diss Dr -Ing , Springer Verlag, Berlin U A
15 Syan, C S (1988) A Computer-Based Surface Coating Selection Methodology for Advising the Design Engineer PhD thesis, University of Hull, UK

CHAPTER 8

Rapid prototyping processes of physical parts

P. M. Dickens

8.1 INTRODUCTION

Many companies are now under intense pressure to reduce the lead time for new product introductions. This is especially so in a market where the lifetime of a product is very short, such as in the computer and electronics sector. DEC claim that they now have a window of opportunity of less than eight months to go from a product concept to launching the product [1]. It is also said that manufacturers of mobile telephone handsets make the vast majority, if not all, their profit in the first six months of sales. With examples like these it is easy to see why companies are introducing concurrent engineering into their management practices to reduce lead times. The most important technological tools to help concurrent engineering are the use of three-dimensional computer-aided design (3D CAD) and the various rapid prototyping techniques.

8.2 COMPUTER-AIDED DESIGN (CAD)

CAD has enabled designers to fully define the shape of objects. With the use of computer numerical control (CNC) machine tools, these can be produced with great accuracy. However, the programming of these machines can be time consuming and they are generally more suited to the machining of complex external surfaces where tool access is obtainable. There are many items with complex internal geometries and small features that render machining either very difficult or impossible.

These problems can now be overcome by using the rapid prototyping techniques. With these, it is possible to use the geometry created during surface modelling or solid modelling. The designs are transformed into a physical three-dimensional model by adding layers of material one on top of another until the complete part is built. Three-dimensional wire frame models are not suitable as they do not fully define all the geometry. Surface models are given a skin thickness as appropriate and the edges of the skins fully trimmed. Essentially, the surface model must be completely watertight and the volume of the part unambiguous. It is because of the difficulty in obtaining watertight models that solid modelling is generally preferred in the world of rapid prototyping.

To be able to transfer the CAD data to the rapid prototyping machine the model is tessellated so that the surfaces are covered with a mass of triangles. The sizes of these depend on the curvature of the surface and the resolution required. The model is therefore approximated even before it reaches the rapid prototyping machine. It is possible to use a fine resolution to give a smoother surface, but the number of triangles then increases dramatically and therefore so does the file size.

These files have become known as STL files and are usually in the 10–20 megabyte range. Obviously, a large part of $500 \times 500 \times 500$ mm cubic capacity of most rapid prototyping machines and necessary required resolution involves transferring large amounts of data. These STL files can now be obtained from most of the advanced 3D CAD software.

Problems are created in using the STL file format due to the faceting of smooth CAD models and dealing with the transfer of large files. It is therefore likely that a new format, known as a slice format (SLC), will become widely used within the next 1–2 years. This is likely to be based on slicing the CAD model within the CAD system and then transferring a series of HPGL files as used on a standard plotter. Companies considering buying a new 3D CAD system would be wise to ask suppliers what plans they have for introducing new file formats such as SLC. Technology in this area is advancing rapidly and it is possible to be left with an out-of-date system in 12 months!

8.3   RAPID PROTOTYPING TECHNIQUES

8.3.1   Stereolithography

By far the most widely used rapid prototyping technique is stereolithography. The main vendor for this technology is 3D Systems, with 90% of machine sales. Although this technique has only been available since 1987 they have already sold over 350 machines, despite their very high cost.

The stereolithography technique is based upon the use of a light-sensitive polymer fluid and sophisticated laser scanning. However, the actual concept is simple: in the 3D Systems machines a vat contains the polymer fluid and within this there is a platen that can rise and fall. The platen moves until just below the surface, leaving a thin layer of fluid. The CAD model has previously been cut into many horizontal slices 0.125–0.75 mm thick (see Figure 8.1). The laser traces out the shape of the first

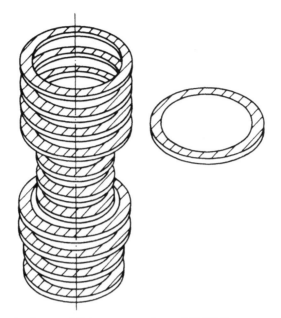

Fig. 8.1 Layer-by-layer additive process from 3-D CAD.

slice on the thin film of polymer fluid (see Figure 8.2). Wherever the laser hits the polymer fluid it sets and becomes a solid. Once the first layer is complete the platen can move down to enable another layer of fluid to form on top of the solidified skin, and so the process is repeated.

Stereolithography can produce highly complicated parts, some of which would be impossible to achieve by conventional manufacturing methods. It is often an uneconomical process for simple straightforward

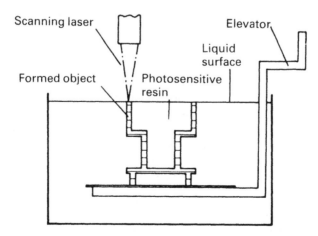

Fig. 8.2 Stereolithography apparatus and process.

Table 8.1   Rapid prototyping system details – 3D systems

System supplier	3D systems Inc			
Process	Stereolithography			
Machine	SLA 190	SLA-250	SLA-400	SLA-500
Max product size (X, Y, Z)(mm)	190 × 190 × 250	250 × 250 × 250	500 × 500 × 250	500 × 500 × 610
Min slice thickness (mm)	0 125	0 064	0 125	0 125
Max slice thickness (mm)	0 762	0 762	0 762	0 762
Product materials	Nine different resins available			
Time to build part of average complexity 150 mm cube (min slice)	Not applicable	Approx 48 h	Approx 25 h	Approx 20 h
Typical material cost	10 pence gram			
Normal accuracy obtainable	90% of dimensions better than +/− 0 1 mm			
Machine size width depth height (m)	1 25 × 0 69 × 1 65	125 × 0 69 × 1 65	1 83 × 3 45 × 2 0	1 83 × 3 45 × 2 0
Machine weight (kg)	270	295	935	935
Power supply	Single phase 240 V, 8 A	Single phase 240 V, 8 A	Single phase 240 V, 30 A	3 phase 63 A @ 415 V
Machine cost	£95 000	£175 000	£298 400	£346 000
Maintenance cost	£8000	£12 000	£30 000	£30 000
M/cs sold to May 1993	Total approx 360			
Typical companies using models from this machine	Aerospace, automotive, electronics packaging, medical			
Parts warranty	12 months excluding laser			
Labour warranty	12 months			
Machine delivery	3 months			
UK importer	3D systems Inc Ltd 0442 66699			
Ancillary equipment required	Slice computer and software Resin stripper and post-curing oven			

parts, as these can be produced quicker and cheaper on CNC machinery. Part accuracy depends to a large extent on the experience of the machine operator because it is possible for parts to distort. This distortion can be overcome by reorienting the part or by building internal or external support structures that are trimmed off later. 3D Systems supply four different sized machines (see Table 8.1) – the smallest is mostly used for small, thin-walled parts whereas the larger machine can produce complete engine blocks!

Part accuracy is an area of much controversy in this technology but people using these machines usually claim to be able to produce models to within ± 0.25 mm and often to within 0.1 mm. In the UK there were only five stereolithography machines in action at the end of 1992, with two medium-sized (250 × 250 × 250 mm) machines in British Aerospace and one large machine (500 × 500 × 610 mm) in Rover. The other two medium-sized machines were in independent model-making bureaus – Formation Engineering Services Ltd. in Gloucester and Laser Integrated Prototypes in March. However, despite the depths of the current recession, large machines have been sold to Rolls-Royce at Bristol, and to the Defence Research Agency at Farnborough. A medium-sized machine has also been sold to another bureau, IMI Rapid Prototyping.

Table 8.2 Rapid prototyping system details – EOS GmbH

System supplier		EOS GmbH	
Process		Stereolithography	
Machine	STEREOS 300	STEREOS 400	STEREOS 600
Max product size (X, Y, Z)(mm)	300 × 300 × 250	400 × 400 × 300	600 × 600 × 400
Min slice thickness (mm)		0 1 mm	
Max slice thickness (mm)		0 4 mm	
Product materials		SOMOS 2100, 3100, 4100 and 5100 Resins	
Time to build part of average complexity 150 mm cube (min slice)		12 h	
Typical material cost		£118/kg	
Normal accuracy obtainable		up to 25 mm ±0 1 mm, above 25 mm ±0 2 mm	
Machine size width, depth, height (m)		1 8 × 2 0 × 2 2	
Machine weight (kg)	800	800	1300
Power supply	220 V, 32 A	220 V, 32 A	220 V, 32 A/380 V
Machine cost	£210 000	£310 000	£415 000
Maintenance cost	£11 000	£11 000	£19 000
M/cs sold to May 1993		16	
Typical companies using models from this machine		Automotive, aerospace, electronics and medical	
Parts warranty		12 months (including laser)	
Labour warranty		12 months	
Machine delivery		12 to 16 weeks	
UK importer		HAHN & KOLB (GB) Ltd 0788 577288	
Ancillary equipment required		Resin stripper	

Rover are recognized as one of the most advanced users of stereolithography. They have many examples where using these techniques have saved them weeks in lead time on product development and many thousands of pounds on model-making costs. The extent of these savings can be seen by the fact that the stereolithography facility costing over £500 000 has paid for itself in less than 18 months!

Due to the limited mechanical properties of the stereolithography resins, post-processing techniques are often used to obtain parts that are suitable for fully functional testing. The techniques most often used are vacuum casting of polyurethane resins and investment casting to obtain metal parts.

3D Systems are now seeing competition in the stereolithography market from other companies in Japan, Germany and France. Their main rival in Europe is EOS, who have a machine with removable resin vats enabling quick changeover times from one resin to another (see Table 8.2). As more resins have become available this has proven to be a very useful feature. It will probably be a standard feature on all systems in the future. EOS market a laser scanner so they can scan an object formed by conventional hand-modelling techniques. They can then produce the object in CAD format and if necessary transfer the information to the stereolithography machine to produce a plastic part.

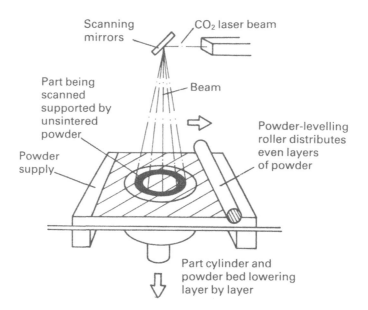

Fig. 8.3 The selective laser sintering (SLS) process.

8.3.2 Selective laser sintering

An alternative to stereolithography is a technique with an even longer name, selective laser sintering (SLS), marketed by DTM in Austin, Texas with a marketing arm in Neuss, Germany. Again this system takes CAD format with the part in slices, but instead of using a polymer fluid powder is used. The SLS machine lays down a thin layer of polymer powder and the laser sinters the powder as it traces out the section (see Figure 8.3). After the section is complete another layer of powder is added and this is sintered to the layer below and so on. The powder that remains unaffected by the laser acts as a support for the sintered part and is brushed off when the part is complete. A range of polymers such as polycarbonate, nylon or investment casting waxes are available for this technique. Using wax models it is possible to produce metal parts using investment casting. The SLS machine is larger than the stereolithography machine for the same working envelope, and more expensive, with part accuracies claimed at $\pm 0.1 - \pm 0.2$ mm (see Table 8.3).

During the sintering process the powders join, but full density is not achieved. Mechanical properties achieved cannot be compared equally with those achieved in normal polymer processes. A research group at the University of Austin, Texas, led by Professor J. J. Beaman is currently working on sintering ceramics and metals using this technique. These materials are a few years from the market yet.

Table 8.3 Rapid prototyping system details – DTM

System supplier	*DTM corporation*
Process	Selective laser sintering
Machine	SINTERSTATION™ 2000
Max. product size (X,Y,Z)(mm)	305 × 381
Min. slice thickness (mm)	0.076
Max. slice thickness (mm)	0.508
Product materials	Polycarbonate, nylon 11, investment casting wax
Time to build part of average complexity 150 mm cube (min. slice)	15–20 h with 0.127 mm, 6–8 h with 0.508 mm in polycarbonate
Typical material cost	DM 100/kg
Normal accuracy obtainable	±0.125 mm
Machine size width, depth, height (m)	2.9 × 1.4 × 1.9
Machine weight (kg)	Approx. 2000 kg
Power supply	230 V/35 A 3 Phase
Machine cost	DM 590 000–890 000
Maintenance cost	13% of machine cost (including materials)
Number of machines sold to May 1993	5 used by DTM Bureau in Texas, 5 Beta machines and 10 Production machines
Typical companies using models from this machine	Automotive, aerospace, medical, computer, electrical, casting etc.
Parts warranty	12 months
Labour warranty	12 months
Machine delivery	1993
UK importer	Europe – DTM GmbH, Dusseldorf
Ancillary equipment required	Included in machine price

8.3.3   Fused deposition modelling

Another technique is the fused deposition modeller marketed by Stratasys. This technique resembles someone icing a cake and building up a 3D structure. The machine incorporates a small extrusion head which extrudes either molten plastic similar to nylon or investment casting wax out of a fine nozzle (see Figure 8.4). The plastic or wax is added layer after layer until the final shape is achieved. This gives models with much higher density than that found with SLS and accuracy of ±0.13 mm (see Table 8.4). This is one of the smaller rapid prototyping machines. It can be used in a design office without the need to worry about handling liquid polymers or powders as in other systems.

8.3.4   Solider

The largest machine available so far is the Solider machine, available from Cubital, an Israeli-based company and marketed in the UK by Sherbrook

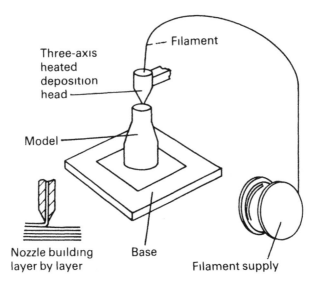

Fig. 8.4 Fused deposition modelling

Table 8.4 Rapid prototyping system details – Stratasys Inc

System supplier	*STRATASYS Inc*
Process	Fused deposition modelling
Machine	3D MODELER
Max product size (X,Y,Z)(mm)	229 × 305 × 330
Min slice thickness (mm)	0 025
Max slice thickness (mm)	0 76
Product materials	Machinable wax, nylon, investment casting wax
Time to build part of average complexity 150 mm cube (min slice)	Approx 4 h
Typical material cost	£143–216 per 1000 m spool
Normal accuracy obtainable	±0 127 mm
Machine size width, depth, height (m)	0 76 × 0 91 × 1 83
Machine weight (kg)	340
Power supply	110 V single phase 15 amps
Machine cost	$206 760
Maintenance cost	$7640 + travel
Number of machines sold to May 1993	More than 20
Typical companies using models from this machine	Aerospace, electronics, consumer, automotive, medical
Parts warranty	90 days or 1 year
Labour warranty	90 days or 1 year
Machine delivery	60–90 days
UK importer	Laser Lines (0295) 267755
Ancillary equipment required	SGI or HP workstation, software and modeller included in above price (typical material cost)

Table 8.5 Rapid prototyping system details – Cubital

System supplier	Cubital Ltd.
Process	Solid ground curing
Machine	Solider 5600 System
Max. product size (X,Y,Z)(mm)	500 × 350 × 500
Min. slice thickness (mm)	0.15
Max. slice thickness (mm)	0.2
Product materials	Photopolymer resin and wax
Time to build part of average complexity 150 mm cube (min. slice)	20 h (in same time system will produce six parts of this size)
Typical material cost	£35/kg resin, £10/kg wax
Normal accuracy obtainable	0.1–0.5 mm
Machine size width, depth, height (m)	4.15 × 1.65 × 2.5
Machine weight (kg)	4 tons
Power supply	3 phase
Machine cost	DM 950 000 FOB ex-works
Maintenance cost	14%
Number of machines sold to May 1993	17
Typical companies using models from this machine	Automotive, bureaus, aerospace, plastics, foundries
Parts warranty	6 months
Labour warranty	6 months
Machine delivery	3 months
UK importer	Sherbrook Automotive 0543 257131
Ancillary equipment required	80 psi air supply, dewaxing unit and oven

Automotive. It is really a combination of stereolithography, fused deposition and a CNC milling machine! Instead of solidifying the polymer fluid with a laser dot a mask of the whole section is used, which is placed over the fluid. Exposed fluid is then flooded with UV light to solidify the polymer. All the remaining fluid is removed by a special vacuum cleaner and molten wax is poured around the solid polymer. When the wax is set a milling cutter removes all the excess wax and mills the section to the correct thickness. This leaves the solidified section of polymer surrounded by wax to provide support. A further film of polymer fluid is placed on top and a new mask used and so on. The mask is produced using a technique similar to that found in photocopying machines, so each mask is transient. When all the slices have been completed, the wax is melted off leaving the complete polymer part (see Table 8.5).

8.3.5 Laminated object manufacturing

The final technique currently available is laminated object manufacturing, available from Helisys in Los Angeles and marketed through

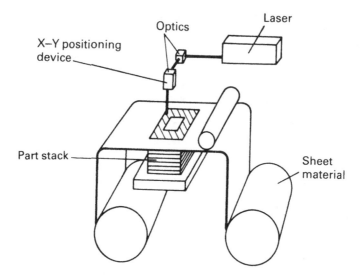

Fig. 8.5 Laminated object manufacturing.

UMAK in Birmingham. This is quite a different approach, as the object is built up with sheets of material such as paper or plastic. The sheets are stacked on top of each other, and adhere to one another.

After each new adhesive-coated sheet is placed in position with a heated roller, a laser cuts out the outline of the part and then crosshatches the unwanted material (see Figure 8.5). Eventually a solid cubic block of sheets of paper or plastic is obtained with each layer glued to the next. The unwanted material is removed by hand and breaks away in discrete lumps, almost like disassembling a three-dimensional jigsaw. Helisys are currently investigating the use of pre-preg sheeting in this system, which would give parts with a much higher strength than any other system available so far. Paper models can be used in investment casting as one material has an ash content as low as 0.2% (see Table 8.6).

The cost of these different machines is still very high, with the cheapest being £140 000 and the most expensive nearer £500 000. It is likely that these prices will fall over the next few years. The companies will manage to retrieve their research costs and other machines will enter the market that are much cheaper and simpler but are just as effective.

8.4 CASE STUDIES

8.4.1 Porsche

There are many examples where companies have used rapid prototyping to reduce product lead times and have very often achieved significant cost

Table 8.6 Rapid prototyping system details – Helisys

System supplier	Helisys	
Process	Laminated object manufacturing	
Machine	LOM-1015	LOM-2030
Max product size (X,Y,Z)(mm)	368 × 250 × 355	810 × 550 × 500
Min slice thickness (mm)	0 05	0 05
Max slice thickness (mm)	0 5	0 5
Product materials	Adhesive coated sheets of paper	
Time to build part of average complexity 150 mm cube (min slice)	18 h	12 h
Typical material cost	£20–75 per solid ft³	£20 per solid ft³
Normal accuracy obtainable	±0 25 mm	±0 25 mm
Machine size width, depth, height (m)	1 12 × 0 76 × 1 14	2 06 × 1 52 × 1 45
Machine weight (kg)	318	1270
Power supply	240 V, 13 A	240 V, 13 A
Machine cost	£115 000	£220 000
Maintenance cost	£7500	£7500
Number of machines sold to May 1993	7	21
Typical companies using models from this machine	Aerospace automotive, medical, foundry	
Parts warranty	1 year	
Labour warranty	1 year	
Machine delivery	3–4 months	
UK importer	UMAK 021 766 8844	
Ancillary equipment required	Outside exhaust, chiller and exhaust	

savings simultaneously. In a recent example, Porsche used selective laser sintering to produce a prototype of a cylinder head. Using conventional techniques of CAD and NC machining to produce sand-casting patterns it would have cost $74 000 to produce the tooling and then $3500 to cast five parts. With rapid prototyping it cost $8000 to do the extra CAD work, $15 000 to produce the wax parts using selective laser sintering and then $3500 to cast them [2]. The time using rapid prototyping was less than four weeks instead of sixteen weeks!

8.4.2 Sunstrand

Sunstrand Electric Power Systems have introduced a concurrent engineering environment with rapid prototyping as a key tool [3]. This has led to significant improvements in several areas. They have found the following benefits, shown in Table 8.7 below. Sunstrand studied benefits in more detail on the design and development of a new current transformer/electromagnetic interference (CT/EMI) module. They compared this with the previous model and found the following results (Table 8.8). It is interesting to note that because Sunstrand could make prototypes much more quickly than previously they could go through several design iterations. This resulted in a product that was easier to produce and had much better reliability.

Table 8.7

Components	Previous system	Using CE and RP
Sheet metal parts	2 weeks	1–3 days
Printed wiring boards	4–6 weeks	2 weeks
Investment castings	12–16 weeks	5 weeks
Machined plastic parts	2 weeks	<1 week

Table 8.8

	Original design	New design with CE and RP
Design cycle time	7 months	4 months
Design cost factor	1.8	1
Engineering change notices	32	14
Production cycle time	35 hours	18 hours
Mean time between failure	1200 hours	8000 hours

Table 8.9

Activity	Previous time %	CE and RP time %
Preliminary design	5	20
Detail design	25	10
Build/test process	50	15
Documentation	20	5
Time saving	0	50

8.4.3 3M

Another company to use concurrent engineering with rapid prototyping is 3M. They have set an ambitious goal of reducing all new product introduction cycles by 50% before 1995 [4]. With the initial projects they have achieved the following results (Table 8.9). With the new design environment much more effort is put into the very early design stages but this is more than paid for by savings in subsequent activities.

REFERENCES

1. Cassista, A. (1992) *What's Important: People – Machines – Technology*. 4th International Conference on Desktop Manufacturing: Compressing Lead Time, 24–25 September 1992, Management Roundtable, Milpitas, California.

2. Smith-Moritz, G. (ed) (1993) *Rapid Prototyping Report*, **3** (5). CAD/CAM Publisher Incorporated, San Diego, USA
3. Gee, Richard. W. (1993) *Case Studies in Rapid Prototyping at Sunstrand Electric Power Systems*. Proceedings of SME Conference – Rapid Prototyping and Manufacturing '93, 11–13 May 1993, Hyatt Regency Hotel, Dearborn, Michigan.
4. Hartfel, Margaret A., Nechrebecki, Dave G. and Scanlan, Michael W. (1993) *Desktop Manufacturing at 3M*. Proceedings of SME Conference – Rapid Prototyping and Manufacturing '93, 11–13 May 1993, Hyatt Regency Hotel, Dearborn, Michigan.

CHAPTER 9

Rapid prototyping of software and hybrid systems

A. Grazebrook

9.1 INTRODUCTION

Software rapid prototyping involves the use of software to create a working model of a system. The main purpose of building a prototype is to help people to understand the system while they are defining the requirements, so that they can write the requirements down in the product specification.

In this section we will be looking at techniques for building prototypes in software. These techniques are valid whether the system is ultimately implemented in hardware, software or a mixture of both. Normally the decision about the hardware/software split is not made until the end of the specification process, when the developers know exactly what they are going to build. When an existing system is being re-worked, the hardware/software split is usually fixed, and the functionality often needs to be modified to suit the practical limitations of the existing structure.

System development starts with an analysis of the requirement, usually written in the language of the application. For example, if we consider the requirement for a laser printer, it would be written in terms of the end user and the attributes that the customer would want, such as print quality, paper size, print speed and so on (Figure 9.1).

The specification of the system involves a move towards a description in the language of engineering. Using the same example, the printer would be further described, detailing the required print engine size, the tolerances and pitch required to achieve the print quality. It is at this point that prototyping can be used to help confirm that the specification meets the requirement.

In some cases, the requirements are never written down, but are taken directly from an expert, and described in terms of an acceptable solution to the requirements. The result of this description is a specification.

A design team takes this specification and looks for ways to implement

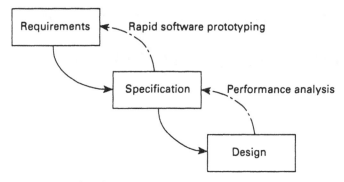

Fig. 9.1   The system development process.

the system given practical constraints: perhaps by using the cheapest manufacturing techniques and the most reliable components. The members of the design team are often of mixed technical backgrounds, each with their own specialization. This can cause its own problems if not properly managed.

This idealized process is very rarely followed through in the way just described. People make mistakes and customers change their minds about the requirements. These two problems account for most of the time and cost overruns in the process of designing systems.

So, designers are often asked to alter the system to meet some new or changed requirements part of the way through the design process. This usually wastes some of the effort spent so far, since the assumptions made at the beginning are no longer appropriate. The designers are put in a position where they can either ignore the new requirement, delay the project, or compromise on the quality of the design by failing to re-assess the original assumptions. Most successful design departments are either geared towards minimizing the cost of late changes, or they become very good at explaining to the marketing department why the changes are impossible.

Prototyping is aimed at reducing design times by giving an accurate specification of the problem at the beginning, and so removing the need for later changes. The technique is successful because it encourages people to think about the details of what they want to build before it becomes too late for them to change their minds. Using prototypes has significant consequences on the lower level design process. Prototypes lead to 'better', more precise specifications, which are less prone to being changed after design has started. This means that there is less need to allow for late changes in the design process.

The technique works by presenting the system in a way that can be more easily understood. The people who are specifying and setting the requirements for the system are left with a better understanding of both the benefits and the disadvantages inherent in the proposed system.

9.2   CONCURRENT ENGINEERING WITH RAPID PROTOTYPING

Concurrent engineering is aimed at improving the time-to-market for products, and also the quality/manufacturability of the design. This is done by pulling the different threads of development together, upfront. The separate development disciplines work in parallel, each on their own development thread, on the assumption that their work will remain coherent so long as they are all working from a common description of the end product. This is inherently a high-risk activity, for the simple reason that you are involving all the different engineering disciplines to the front end of the design-cycle, and are making commitments to manufacturing and purchase much earlier than normal.

In order to make this exercise a success, it is essential that all the people in the team are working together coherently. This means that they must have a common understanding of the problem that they are trying to solve, or to put it another way, they must all be trying to design the same thing. This increases the importance of providing a good, solid specification. If you have already built the production line, it becomes very expensive to change the design.

9.3   AIRCRAFT AVIONICS – AN APPROACH TO SPECIFICATION

Let us look at ways in which Rapid Software Prototyping can be used in practice. This example is based on some work performed at Rockwell International, applied to some avionics systems on a Boeing 747-400 aircraft. They ran a case study based on the development of the onboard maintenance system (OMS) within the aircraft.

The avionics systems in modern commercial airliners are becoming increasingly complex, with more functions being incorporated in the automatic control systems of the aircraft in order to reduce the workload on the flight crew. The development techniques that had been used in the past on smaller systems were not appropriate to these more complex systems, so the developers decided to use rapid software prototyping to validate the requirements of the OMS.

9.3.1   Prototyping process

There were several steps involved in the development and use of the prototype. These are shown in Figure 9.2. The scope of the model was based on the limited amount of resources and personal expertise of the team members. While they considered rapid prototyping very valuable,

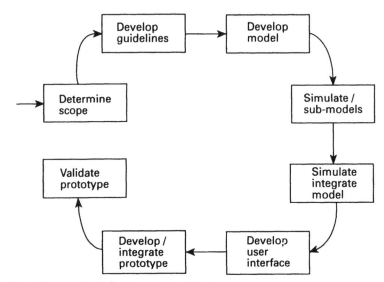

Fig. 9.2   The model definition and validation process.

there was no need to base the model on the complete functionality of the OMS system.

Before any real work was done, the team established guidelines to ensure that the work from each member of the team interfaced properly. These guidelines made clear to each of the members of the team what their responsibilities were, and what format they were going to use to pass information between each other. The guidelines could also be used to form the basis of the future quality assurance procedures. (Remember that this is a case study. Part of the purpose is to understand what should be done for future projects.)

In general, it is important to consider the uses to which the model will be put when deciding what kind of model to build. This will then be reflected in the way in which the model is constructed, which in turn is reflected in the way team members co-operate together.

As should be clear from Figure 9.2, the developers constructed the model of the whole system by making several sub-models of the parts of the system, and then integrating them into a complete system model. The principle of divide and conquer should be used whenever constructing a model of any reasonable size; it allows the individual pieces to be checked separately while they are still small enough to debug properly.

The lesson to be learned from this example is that there is a significant planning stage that comes before committing to build the prototype. The important issues to resolve are:

- What questions will be answered by building the prototype?
- What form should the prototype take?

- How much of a prototype do you really need to build?
- Could you get the same answer more easily?

9.4 TECHNIQUES IN RAPID SOFTWARE PROTOTYPING

The nature of the prototype is heavily dependent on the nature of the system that you are developing. Given that a prototype can't be a mock-up of everything (otherwise it would be the final system), you have to identify the relevant characteristics of the system that you are trying to model. The most common characteristic to model is the user interface. Other possibilities include modelling the behaviour of the system, or the performance of the system.

Figure 9.3 shows a standardized component of the system development process. What it does not show explicitly is the return path or verification path. At all times in system development, the aim is to move from concept through to realization, or, to put it another way, to put ideas into practice.

The objective of the planning phase of the project is to define the steps in the development process. Figure 9.3 can be used as a basic building block in the development plan. For each of the development steps, the method that will be used needs to be defined, and the expected information content of the result of the development step (the improved representation) needs to be described.

There are a variety of standard steps in this process, from requirements analysis to manufacturing. All of them follow the same basic pattern: the initial representation is taken, something is done to it, and a result is

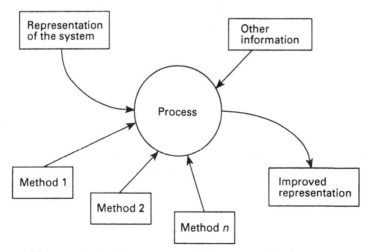

Fig. 9.3 The standardized component of the system development process.

Table 9.1 System development, early stages

Development stage	Information added	Verification	Representation
Requirements analysis	The customer's requirements and ideas. Standards.	Discussions (with modelling/ prototyping).	English text. Diagrams.
Specification	Engineering detail. Definition of development process.	Requirements traceability. Checks on model (prototyping) customer reviews.	Functional model + detailed descriptions. Development plan.
High level design	Architectural description. Choice of implementation technology.	Consistency checks with specification. (Performance analysis.)	Architectural model. Hardware/ software split.

produced, which is hopefully an improved representation of the system. Usually someone checks the improved representation to ensure that it is consistent with the initial representation.

So, what do we need to add to the initial description of the system to improve it? Table 9.1 lists the early stages of system development. A set of stages that are close to those generally used in industry is shown, with added references in parentheses to rapid prototyping techniques. The division of the stages remains a matter of opinion, and none of the stages have a universal definition.

9.4.1   User interface mock-up

The most commonly-used technique in software prototyping is to draw the user interface, and to present a working example of part of the system. A simple example is an electronic point-of-sale terminal (EPOST). An EPOST is a shop cash-till, linked to a computer, which contains the latest pricing and stock control information.

The EPOST prototype used a personal computer that was programmed to go through a few standard sequences. The sequences were chosen to represent the typical transactions that were performed on the machine, e.g. entering a few groceries and closing a credit-card transaction. The computer screen display mimicked the output that would appear on the terminal screen. The prototype stepped through the display sequence when the user pressed the correct key, and beeped if a wrong key was pressed.

Some 'typical user' trials were conducted on the prototype EPOST, in order to find out what kind of use (or abuse) the terminal got when entering the typical week's groceries. The testers tried a few different variations on the sequence of operations and display, and recorded the mistakes as well as the successes of the users. The designers used this data, and the resulting comments from the trials users, to select the best design for the EPOST. The EPOST developers also used the data from the trials to work out loading figures for the central stock-control computer.

This example highlights the main characteristics involved in using rapid prototyping to produce user-interface mock-ups:

- The presentation of the information is as close as possible to the end product.
- The processing of the information is hard programmed or very restricted. This helps to focus the objectives of the user trials as well as saving effort in making the prototype.
- The information gained from testing the prototype is used to improve the user interface, but also can be used in other ways. In this example, it was used to provide information about the stock control computer. It is worth thinking about what other information a prototype can be used to collect, and to build it accordingly.

In actual fact, this technique can be (and often is) applied in other ways without the use of computers. One of the simplest examples involved a member of the marketing department putting up a series of pictures of some sound recording equipment, and explaining to some sample users what happened when different buttons/functions were pressed. There were also some useful side comments about how this machine was different from others on the market. The users' comments and suggestions were used to modify the equipment design to make it easier to use and understand.

To sum up, this kind of prototype is useful for systems where the user interface itself is important, and also where information is needed about the effects/patterns of use.

9.4.2 Behavioural prototypes

For some classes of system, the behaviour is the key aspect, and therefore the characteristic to be examined most carefully. Unlike the user interface prototypes, where pictures of the system and its interface are used to describe the system, the behaviour needs to be described using a specialized notation.

There are several different diagrammatic notations used by designers today to describe system behaviour. The statechart (an extension of state-transition diagrams) is one technique often used in embedded control systems, SDL is used in the field of telecommunications and protocol modelling, coloured petri-nets have been used to describe secure banking transactions. There are also many other notations used in specialist industries.

Let us look at a hypothetical example of a behavioural prototype. The avionics systems for modern jet fighters are notoriously complex and difficult to specify. This is because there is a large amount of interaction between different system components, and the timing aspects of the interaction are no less important than the data content.

The aircraft avionics suite includes a bomb aiming system. This system helps the pilot to navigate to and to aim the munitions at the target. It communicates with the stores management system and interacts directly with the pilot/navigator through the head-up and head-down displays. It also receives information from a variety of other systems, such as the air-data and inertial navigation systems.

The specification for the weapons targeting system was created using a tool called Statemate. The specification was based on a requirements description. The requirements were predominantly written in the English language, supported by a variety of informal diagrams. The English-language description inevitably left out some of the detail that is required in a specification. The development team accepted this, and decided to use simulation and prototyping techniques to fill in the details and resolve ambiguities.

While the developers were entering the model, they used Statemate's simulation capabilities.

1. This confirmed that the model that they were entering had the behaviour that they expected.
2. It allowed them a better understanding of the way that the different aspects of the model worked together.

Throughout this stage, the developers asked the requirements specialists detailed questions about the requirements of the model. These questions were prompted by a need to fill in some of the details that were left out of the requirements document, and also when the simulation showed inconsistencies or ambiguities in the original requirements.

At the same time as the model was created, a separate group of engineers created simplified models of the environment. Some of the models were based on discussions of typical operational use, and some on the direct responses of other, connected, pieces of equipment. This allowed combined simulations to be run, which helped the developers understand the normal behaviour of the system.

The next stage was to present the specification model in a form that could be understood by the requirements specialists. An initial prototype was made using Statemate's panel graphics editor. This editor allows a user interface to be created, with buttons, dials, switches and so on that are bound to the contents of the model. The panel combined with the model is then compiled into a prototype that is controlled using the panel. The panel was presented initially to the requirements specialists, who provided feedback to the developers with suggestions for change to the specification.

Once the obvious problems had been ironed out, the developers took the code generated from the specification and integrated it into a flight simulator. At this point some pilots were brought in and given the opportunity to try out the prototype in some realistic operating scenarios.

When all the problems and suggestions had been resolved, the specification (including the prototype) was baselined, and released as a basis for the design of the system.

This example illustrates two of the common characteristics of nearly all behavioural prototypes.

- The prototype includes not just a model of the system, but also a simplified model of the environment in which the system will operate.
- Some of the effort that goes into using the prototype to get useful results is spent on acquiring suitable test data. Do not under-estimate the importance of this.

9.5 SUMMARY

This chapter is all about information gathering. In it, some of the techniques that are commonly used at the analysis stage of system development have been explained, and some guidance provided in the techniques of software prototyping that are used to gather the information about the system.

Although the principles of software prototyping are common to different kinds of development, the practice of using software prototyping is different for embedded systems and for information systems.

There are relatively few books that are dedicated to the practice of creating software prototypes, but there are a large number of books that discuss the whole process of system or software development and in doing so make a passing reference to prototyping as one of the tools involved in the proposed development cycle.

FURTHER READING

1. Rahmani, A.G., Stone, A.G., Luk, S.M., Sweet, S.M. (1992) Rapid Prototyping Via Automatic Software Code Generation From Formal Specification: A Case Study. In *Behaviour Models Specifying User Expectations* (ed. J.A. Kowal) Prentice Hall Publishers, NJ, USA.

PART THREE

Product Design, Support and Management Tools for Concurrent Engineering

CHAPTER 10

Software tools for the product development process

S. C. Hitchins

10.1 INTRODUCTION

While computer-aided design systems make it possible to build a database of design information, engineering data management software focuses on improving the availability of that information for faster product introduction. The combination of CAE/CAD/CAM and engineering data management systems today has successfully provided increased productivity in different segments of the product life cycle. However, as the capabilities of these systems grow, there is an increasing demand for a greater level of information management for cross-functional operations.

The new emphasis for organizations is on teamwork, information sharing, and timely, informed decision making. These are key components of concurrent engineering. However, most organizations today have multiple management processes and use separate systems to control different aspects of the process. These include engineering data management, release control, configuration management, material requirements planning (MRP), program management and other methods. Each of these systems uses information about components, tasks, and product structure as basic data elements for controlling the specific process. Inevitably, a great deal of effort is necessary in order to reconcile discrepancies between the information in these systems. This is a well-known bottleneck which affects productivity and increases time to market. For this reason, organizations are looking for a single, integrated database to which all existing applications interface, so that changes from one department are visible immediately to others.

10.2 LOGICAL FRAMEWORK

A logical and systematic way of looking at products is in terms of a hierarchical tree that can be used to describe multiple views of assembly

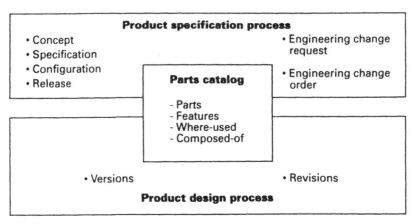

Fig. 10.1 Overall product specification and design process.

or product structure. This hierarchical product structure contributes to the definition of the processes used for design through production; for example, an automotive company is organized into groups that focus on chassis, engine, transmission, body and so on. Each of these organizations uses a different view of the product structure to meet their requirements. In essence, the process is the manner in which our customers design and manufacture products, while product structure is the means used to describe the process.

Because product structure drives the process, there is an increasing demand for configuration management systems that can capture changes at the source without requiring replication of data. In addition, because our customers use dynamic processes, the configuration management system must be able to manage change over time. An approach that leverages data capture in the engineering and design environment is up and down-stream in the product process. Increased integration is desirable. The types of product structure information required within an organization dictate how tools should be developed and used. Three basic models commonly used for encapsulating product structure information are illustrated in Figure 10.1.

1. **Product Specification Process.** This is the process of identifying requirements for new products or changes to existing products, especially with respect to multiple configurations of a product built from the same basic set of designs, such as a product line. General requirements include the following:
   (a) The ability to describe the configuration from combinations of information in the parts catalog.
   (b) Creation of alternate configurations for different markets.
   (c) Identification of design, manufacturing, or market alternatives relative to configurations.

   (d) Specification of the production dates or serial number effectivities.

   (e) Support for an engineering change proposal/order/notice process.

2. **Parts Catalog.** This is a catalog of the products built by an organization. It includes the following information:

   (a) components and the relationships among them;

   (b) parts;

   (c) features.

The information contained in the parts catalog can be used for queries such as 'where-used' and 'composed-of'.

3. **Product Design Process.** The product design process includes creation of a new design or modification of an existing design to implement an engineering change. In most cases, the process begins with a search of the parts catalog to locate any similar designs. For this reason, the product design process requires the ability to do the following:

   (a) Manage assembly information, so that variations of the same item can be supported at the same time.

   (b) Define geometric regions or zones of a product organized by discipline.

Each of these processes may or may not be used in conjunction with the other. Figure 10.2 illustrates the overall flow of processes within an organization, which can be related to the models described above. As shown in Figure 10.2, the goal of the document control process is to bring as much information as possible under the management of the system without reference to individual design projects. This is where the parts catalog fits into an organization's processes. The process management phase builds on the document control process, and is equivalent to the product specification process model. Design projects are created during this phase, and each design process is captured in a defined workflow, which is reflected in the review and approval steps for each project.

   The change management phase extends the design review process by incorporating non-design data and developing an engineering change order (ECO) procedure. This phase corresponds to the product design process model. The final phase manages configuration and data integration throughout the product life cycle. This phase addresses all areas of product development, including the definition of marketing. A discussion of the three models for use of product structure information is provided in the following sections.

10.2.1 Product specification process support

The product specification process is focused on identifying product requirements prior to initiation of the design process. Each specification

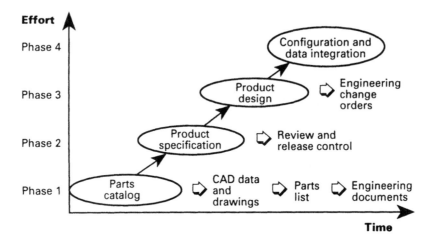

Fig. 10.2 Overall flow of processes in an organization.

has a finite lifetime – this is expressed in terms of the period of time during which the configuration will be manufactured or the unit serial numbers that will be built using that particular configuration. Product specification does not always have to involve the product design process; however, it usually creates a series of specifications for design of new parts or sub-systems, or change proposals to existing systems. These specifications are usually expressed in terms of customer-orientated features such as an electric sun roof, a specific avionics system, or airflow characteristics for an aircraft engine. These features then become engineering assemblies of the set of components that produce the desired product.

This process creates the specification of a product structure before any engineering of functional sub-systems, components, or detailed parts is done. Consequently, the detailed design phase must be managed and co-ordinated by the product specification process. For this reason, it must be possible to integrate any design software that works with product structure information into the customer's product specification process. The general requirements for such a design product include the ability to describe the configuration from combinations of information in the parts catalog, to create alternate configurations, to specify production dates or serial number effectivities and to implement engineering changes. The product specification does not require detailed design information as an integral part of its database; however, it must know where to find this information and how to control access to it.

Because the product specification process involves examination and changes by more than the engineering and product configuration departments, product specification support must also include the controls to allow access and management of assembly information across the entire organization.

10.2.2 Parts catalog support

A parts catalog must provide an industry-standard representation of products that is usable as the basis for creating new product designs from old ones. In addition, this system must support impact analysis for proposed changes – in terms of potential cost, lead time, or customer requirements – by showing in which products current designs are used. The parts catalog must also provide feature-oriented, functional, customer-defined and physical viewpoints of sub-systems of the product. Support for basic product structure, multiple revisions of products and product family information must also be provided.

One of the major uses for a parts catalog is to support standard part libraries – collections of parts that have been authorized for use in designs. In electronics, this is often a set of components from authorized suppliers; in mechanical industries, this is a set of previously approved designs. The principle use of a standard part library is to avoid having to re-design a component when an existing design will satisfy the requirement. Each component is classified according to attributes of interest in the design or manufacturing process, and designers may use any combination of attribute values to search for designs. Parts catalogs are also used to support standard tool libraries.

Parts catalogs are also useful to support multiple views of assembly or product structure. These different views are specific to areas of responsibility or functional areas in a company. For example, an automotive engineer may wish to see a view of the components that make up a specific feature, such as a sun roof, car seat, or instrument panel. A manufacturing engineer wants to see all of the components that are built during a specific stage of the assembly process – for example, all parts in the trunk.

10.2.3 Product design process support

In most cases, the design process starts with retrieval of an existing design from the parts catalog in response to a change request. Even for a new specification, there is usually an analysis of existing designs using the classification capabilities of the parts catalog. In addition, in the course of satisfying a particular requirement, engineering groups often refine designs through comparative analysis with alternative designs – several alternative designs for the same product may be worked on at the same time. In order to support concurrent engineering, these designs must be frozen at various stages of evolution to allow associated groups to examine them. For this reason, product design process support requires a system for managing assembly information that can support alternate in-work versions of the same item.

Alternate design versions also proceed independently throughout the design release process – sometimes the versions are merged prior to release, and sometimes not. The design process proceeds from initiation to release of the design through a series of steps or milestones. Assemblies are tracked in the same way, often with the added constraint that the assembly cannot be approved and released until all of the components are approved. This requires the ability to track in-work versions of the product against milestones.

Organizations with complex design processes for large projects, such as aerospace and shipbuilding, use design techniques that break the project down into sections for manageability. These sections are created by defining generally rectangular regions of the overall product. In shipbuilding, this is often the region between major bulkheads on a single deck. In aircraft design, it is cross sections through the fuselage bounded by decks. Responsibility within a region is given to a design team to design structures, fixtures and hangers, electrical distribution, hydraulics, air conditioning and other sub-systems. This technique arose to assist with the co-ordination of interacting design disciplines by defining a manageable subset of information (the region). The design software must be capable of breaking data into zones and disciplines in order to support these types of project.

Similar requirements exist in manufacturing planning operations across vertical market segments. These operations use a manufacturing technique called zone-by-stage, which seeks to build all of the components in a particular product region as a unit. In consumer electronics, this would be plug-replaceable units, which facilitate manufacturing and repair. The zone-by-stage manufacturing technique is also becoming popular in automotive and other volume-manufacturing operations.

10.3   DEFINING REQUIREMENTS

As previously mentioned, each of the types of product structure support may or may not be used in conjunction with the other. However, these requirements can generally be related to the size and process complexity of the customer. For example, the major requirement of small businesses that use processes of relatively low complexity is for a parts catalog, while larger businesses with highly complex processes require product specification support. Businesses that fall in between these two categories mostly require support for the product design process. The following paragraphs describe the requirements of some example companies who fall into different categories of size and process complexity.

10.3.1   Hurel-Dubois (managing engineering change)

As a leading player in the aircraft industry, Hurel-Dubois is the first to acknowledge its debt to computer-aided design. When it wanted yet more speed and quality, it quickly realized that it had to look beyond CAD/CAM and embrace engineering data management (EDM). 'The capabilities of advanced computer design are spectacular. But if you look beyond them you may find the productivity benefits that they bring will pale into insignificance compared with the next step, which is using your design data to control your engineering development process. For us that is where the priority now lies.'

So says Mike Jones, chief engineer with Hurel-Dubois UK, the Burnley-based manufacturer of high technology aircraft components and structures. HDUK, having enjoyed the benefits that 3D CAD/CAM brought to its design office, has now extended those advantages to its whole engineering and manufacturing process by adopting engineering data management (EDM).

HDUK is part of the French-owned Hurel-Dubois Group, which specializes in components for jet engine nacelles or pods. Its products are now standard equipment on aircraft including the Boeing 747 and 767, with RB211-524 engines and the Airbus A330 with Trent engines. A major part of the company's work is designing and manufacturing thrust reversers. The average thrust reverser consists of around 1600 parts.

Jones describes the problem: 'A typical product can have 1000 parts, plus ten types of fastener, and can incorporate 100 to 1000 modifications over several development units. Then add to this as many as ten different engineering functions all needing the latest information so that they can achieve lead times of 12 to 24 months from receipt of aerodynamic lines to delivery of the first unit.' A daunting task in anyone's book. 'For an engineering company like ours, working with a combination of a high parts count and constant change, getting the product right first time used to be a pipe dream' says Jones. 'In this type of engineering, it was common to say the first units were automatically the worst units. But your customer's opinion is formed by those first units. Here at Hurrel-Dubois, we always prided ourselves on our ability to manage change. But developing a 1600-part product at speed used to mean compromise. So, you might still find problems occurring by lot 100. Of course, 80% of cost is locked into the first 10% of design time. Getting it right early can make a spectacular difference, particularly in a highly competitive sector like ours. So, we decided we had to be able to offer our customers both unequalled speed and quality.' Concurrent engineering, where different disciplines work as parts of an integrated project team (rather than each waiting for its predecessors to finish with its stage and pass it on) was seen as one way of compacting the process.

But going for the productivity benefits of concurrent engineering when you are working with huge amounts of data like Hurel-Dubois can leave an engineering operation frustrated by an inability to ensure everyone is working to the same data. This is why some firms still stick to the slow but orderly handovers of sequential engineering, and remain bogged down by the need for constant and expensive re-work. Not for them the faster-moving waters of concurrent engineering, for fear of not being able to cope with waves of changes all coming at once.

Jones found he was getting similar comments from people frustrated when the speed of the product development was pushed ever higher. The first was: 'Engineering has introduced another change and we're still waiting for drawings. And we are short of essential information that we need to finish the job.' Another comment was: 'The first hard facts we get about the new product don't arrive until first scheme release, when we find the design content is not achievable within our lead times or our allocated costs.'

Jones continued: 'In recent years, we had achieved major quality improvements with the help of Computervision's CADDS computer-aided design system. Geometry control and co-ordination of components, shape-making tools, manufacturing processes and assembly fixtures had all improved greatly. So further improvements in these areas of the development process were less of a priority and harder to achieve.'

(a) Essential CAE

Last year, the company decided it had reached the stage where computer-aided engineering (CAE) was essential to generate the product definition. 'What we wanted, in effect, was to enable the controls which were applied to sequential engineering to be maintained whilst achieving a high level of concurrent engineering.' However, like many firms before, Hurel-Dubois discovered its search for the best of both worlds was in danger of taking on the character of the quest for the Holy Grail.

'The other data management packages on the market simply did not address the needs of controlled concurrent engineering and the design-related methods of configuration control typical of the aerospace industry.' Considerable progress was made with the introduction of CV's Concurrent Assembly Mock-up and EDM version 4.2. But it was not until Computervision offered the company the early opportunity to combine Concurrent Assembly Mock-up with the new functionality of EDM version 5 that Hurel-Dubois felt it had a system that potentially had all the answers Steve Lees, Jones's engineering and development manager responsible for implementation, takes up the story: 'Concurrent Assembly Mock-up defines the product structure in the form of a family tree, while EDM automatically provides the parts lists, bills-of-material and concurrent information. Essentially, we are using a tailored implementation of the two products working together, generated by a team drawn from CV and HDUK. It has allowed us to move from being a function-centred organization to a product-centred one.'

(b) Typical philosophy

The approach is becoming increasingly typical of CV's philosophy for meeting manufacturing needs in the nineties, through working closely with customers to adapt products to their exact needs. 'That's what attracted us to the CV solution – its ability to offer a standard package which was customized by a joint team based on an understanding of our needs.' The benefits are already making themselves felt. Reports are now provided from a single database. This includes, where necessary, transposition into HDUK's MAAPICS MRP system.

With EDM 5, Hurel-Dubois can for the first time feed partial bills-of-materials representing engineering build standard into its MRP system. 'Previously, it just couldn't deal with it. EDM 5 gives us the vital ability to track a component's history throughout the development process.' Re-work has been cut considerably and, along the way, the company has also been able to achieve a significant reduction in development stock. 'But what EDM means overall is that everyone is working to the same concurrent engineering database; they're all singing to the same song-sheet. Every step of the process, the slightest change, is authorized and everyone is instantly updated.'

By allowing Hurel-Dubois to control change so tightly, EDM 5 will give HDUK customers the advantage of continuing to define and enhance their product well into the development cycle, knowing that HDUK has the built-in flexibility to respond. As Lee says: 'We've turned change control from what was an engineering function into what is now a business function.' Lee, the implementer, is quick to point out that effective concurrent engineering requires changes in organizational methods, which tools such as EDM can then accelerate: 'We were keen not to graft a big investment onto existing, inappropriate practices, but to change them first.' The evidence of this is Hurel-Dubois' new open-plan drawing office, where the walls have come down to allow teams from different disciplines to work together and exchange information freely. There is a regular exchange of staff between design and shop floor to enhance teamwork. Adds Jones:

'The layout of the drawing office was worked out by users, and it overcomes the old sequential problems where design never talked to the detailing office.' Today, Hurel-Dubois is well on course to finish the development of its first major new product using EDM-controlled concurrent engineering from start to finish. Its nose wheel undercarriage doors were due for completion by May 1993. Speaking at a recent Economist European Manufacturing Forum, Computervision's UK managing director Garreth Evans pinpointed the adoption of EDM as pivotal in honing manufacturing's competitive edge: 'In the sixties, cost was the battle cry, in the seventies it was quality, in the eighties, time to market and in the early nineties, niche marketing is seen as the route to competing successfully in the new global marketplace. By the year 2000, the battleground will have moved onto exploiting the database. And that means managing engineering data effectively.' 'The year 2000?' queries Jones, sitting back in his chair in Burnley. 'If anybody out there leaves it till then, they'll be far too late. In fact, they probably won't be around.'

10.3.2   Short Brothers: from wood and canvas to world class

Short Brothers traces its remarkable history in aviation from the first order for six 'Flyers' placed by the Wright Brothers in 1908. Today, the company is a world-class player in the global aerospace business, making extensive and highly successful use of CAD/CAM.

Short Brothers likes to be first in what it does. From making six 'Flyers' for the Wright Brothers in 1908, to the design and building of the world's first vertical take-off and landing aircraft in the fifties, the company has a tradition of innovation. In fact, Dennis Barritt, CAD/CAM manager at the Northern Ireland firm, still uses the original drawings of the Wright Flyers in his CAD presentations.

The drawings have been rasterized and are held in the company's CADDS system, where they nestle digitally, if a little incongruously, within a database that now includes a full-blown electronic mock-up of a modern aircraft wing. Becoming one of the first aircraft manufacturers to produce a digital mock-up of an assembly as complex as a complete wing on CAD is Short's latest step in the path of innovation.

It is a road that has carried the company, which now employs 9000 people, from working in wood and canvas to its place as a major player in the global aerospace business. Short's route to its pioneering wing mock-up was by way of a much simpler project – an engine fan cowl door. This was one of several early projects on which Barritt's CAD/CAM team cut its teeth before applying the design-for-build lessons to a whole wing. Barritt explains:

'We moved into CAD in 1981, but for eight years we used it only as an electronic drawing board. But in 1989, we saw the potential for using CAD/CAM to close the gap between design and manufacture, and I took on the job of heading up a small team to drive that integration. Any firm managing this kind of change needs an integration champion with firm senior management backing to cut through the traditional barriers. This is particularly true in a large company where departments are natural ways of segmenting an otherwise unwieldy organization.'

Barritt's approach is one that is becoming the tried-and-trusted way of easing concurrent engineering methods into traditional, serial manufacturing. By starting small, on single projects to prove the technology and working methods, Shorts managed its growing use of leading-edge technology and convinced middle management.

'It's the comfort factor that's important', says Barritt wryly. 'You move forward in three-monthly steps, with a presentation at the end of each step so you can appreciate what you've achieved. What we did was to set up design-for-build teams to work on particular projects, so that they sat with the designer and had an input. Our approach was to take the threads of CAD/CAM and weave them into the whole design-to-manufacturing process. With the fan cowl door, this involved analysing how we had designed and manufactured composite components in the past and then starting again, but this time concurrently with CAD-to-CAM links.'

The fan cowl door is part of a nacelle or jet engine pod. As aircraft have become more complex over the years, a major shift in the industry has been to divide manufacture. Large assemblies are designed and manufactured by major sub-contractors across the world and then fitted together by the airframe manufacturer. One of Short's particular skills is nacelles and nacelle components, in which it has pioneered the use of weight-saving composites. The carbon fibre and Kevlar composites are expensive to work with, not least because of the nesting difficulties inherent in working up the doubly curved shapes of nacelles.

'With manual methods, we had been achieving 45–48% materials utilization, just like the rest of the industry. There were also built-in time delays, since we had to wait for the lay-up tool to be made and then use that tool to size the materials by literally draping the composites over it. We then cut around the edges and threw away the waste.'

Barritt's team progressed the door design on-screen, through stress and weight definitions, and analysis to detailed design of fittings and definitions of the composite plies to be used. At the same time – and with access to the same data – the tool design office was developing detailed design of the composite lay-up tools and using CAD data to generate NC cutting paths. Production engineering, again using the same data, ran a flat pattern development and nesting programme.

By the time the composite ply had been nested, the lay-up tool had been machined and was ready to go. Says Barritt: 'By the end of the project, we had increased materials utilization to 85%, bringing an annual saving of around £400 000. And lead times had been cut by 30%.' Many of these early lessons were carried over to the much more complex task of designing and manufacturing a complete wing. 'Some of the processes we derived from work on the earlier project. The chief difference, however, is that with the door you could check interference physically by putting it into a nacelle.

'With the wing, you are given the aerodynamic shape and the job is to sandwich in the components, ensuring that there are no clashes. The only way to avoid constant reiteration is through digital pre-assembly. And this was the great leap forward.' Central to the CAD/CAM team's success was the creation of an intelligent router that semi-automatically chooses the optimum piping and electrical paths on-screen, carrying out automatic hard and soft clash detection on the way.

Barritt regards the wing mock-up as the high point of his team's achievement so far – pulling together the threads of CAD/CAM, which have seen parallel improvements in lead times from composites to sheet metal, machined items and piping. 'With sheet metal work, our lead times are down just as much as in composite work, from four weeks on average to four hours. On piping, we're down from one week to three hours.'

Short's investment in its 130 CADDS workstations would not have got it so far, however, without an equivalent commitment on the part of its Canadian parent company Bombardier to invest in state-of-the-art NC machine tools to handle the CAD/CAM data. As for the next frontiers, Barritt sees Shorts moving towards knowledge-based systems, with parametrics eventually slicing further into lead times. Short's solid modelling capability, which is enhanced in CADDS 5, and the continuing use of composites, will allow larger and hence fewer components to be used, making the process still faster. Says Barritt:

'From 1909 to 1979, nothing changed – it was a 70-year era of paper drawings. From 1979 to 1984, we gradually replaced paper with CAD as an electronic drawing board. From 1984 to 1989, we massaged the system into a true CAD/CAM package. And over the past three years, we have discovered that the only real limit on what we can do with a system like this is our imagination and vision.'

10.4   OTHER CASE STUDIES

10.4.1   Rediffusion, Inc. (product specification process support)

Rediffusion, Inc. is a medium-sized company that produces extremely sophisticated aircraft cockpit and control tower simulators. Their product process is extremely complex and uses leading-edge technology. All of Rediffusion's simulators are custom built to meet each customer's requirements; however, they estimate that 50% of their design is created from common elements, 30% from elements associated with a product type, and only 20% based on customer-specific elements. Their goal is to create a complete catalog of simulator design elements and product options, and to be able to select a particular configuration from the options within the catalog.

Currently, Rediffusion lacks the capability to duplicate the specifications of a previously-built simulator without actually going to the customer site and 'reverse engineering' the specification. This indicates a clear requirement for a parts catalog; in addition, they want to address configuration management issues once the parts catalog is in place. Their plan is to create the parts catalog, extend it to support their bidding process and then incorporate their product specification process.

10.4.2   GEAE (product specification process support)

General Electric Aircraft Engines (GEAE) is an example of a large business with a highly complex product process. GEAE requires integrated product information that spans existing designs (parts catalog), with support for assembly information co-ordinated with design information in the design process (product design process support). GEAE also requires configuration management, with tracking of design derivations, alternate assemblies and as-built versus as-designed structures (product specification process support).

10.4.3   AMP (parts catalog support)

AMP is a manufacturer of electronic connectors, cables and cable harnesses which is organized into both moderately large and very small design and manufacturing organizations. Their product process is relatively simple, and the dominant requirement is for a parts catalog that supports where-used queries and ECO processing. This is because AMP frequently finds the same part is being manufactured under two different part numbers, or that two different parts have the same number. One of AMP's biggest problems is identifying where a particular part is used in other assemblies in order to determine the impact of a proposed change. AMP has already begun to use the concept of a family of products, but does not currently use configuration management on the assemblies. Their goal is to first implement a parts catalog to manage product information, then eventually move into configuration management. Currently, AMP is using MEDUSA and Unigraphics systems managed by EDM.

10.4.4   Philips (product design process support)

Philips Consumer Electronics is organized as a family of co-operating organizations; some large, and some very small. Their product processes are moderately complex to fairly simple, and require support for complex, shared product structures with alternatives and versions. Philips uses the concept of 'product families', and requirements for product structure management are focused on the product-design process environment. To support this, Philips requires the ability to create a product classification scheme that groups products made from similar technologies (parts catalog). The use of product families helps control component diversity, reduces parts lists, gives a better overview of future material requirements and helps to plan commercial diversity without having to redo PCB designs. Philips also wants to change its internal design documentation from product-based to product-family based.

In addition, Philips has a problem with data volume and inconsistency, and wants to implement the product-family structure concept to assure consistency of data across company functions. They require access control by zone and by part – a given zone must be usable within different members within a family. Their design process is to define commercial requirements, break it down by sub-function, and relate the sub-functions to markets. Philips then defines the product family's architecture, mapping sub-functions onto the design building blocks to create variations of the design building blocks for each target market. Once this is done, Philips lists all the assemblies for the family as a whole and the generic building blocks for each assembly. To create a particular product, they choose the correct version of the correct variation of each building block for that product's family.

10.5   COMPUTERVISION PRODUCTS

In order to provide the necessary products to support these require-ments, Computervision's data management strategy is focused on providing an integrated suite of software products that enable our customers to manage product structure information. We recognize the requirement that the product-structure management system be an industry standard, full-featured system that supports parts catalogs with where-used impact analysis, product design and engineering processes with version histories, and product specification processes for engineer-ing change orders and configurations. Computervision provides a suite of products that support integrated information management on several levels, as follows:

- **assembly level** – Concurrent Assembly Mock-Up
- **attribute level** – CADDSInformation
- **zone/discipline level** – CADDSControl
- **file level** – EDM.

Figure 10.3 illustrates how these products fit together. As can be seen in the illustration, EDM can also manage more than just CADDS-generated data – it can manage virtually any kind of data (e.g. MEDUSA, Unigraphics, DIMENSION III, Catia). In addition, Computervision is currently developing a high-end bill of material and configuration management database manager called EDMProductManager (EDMPM).

CADDS provides a single, integrated database to which existing applications interface, so that changes from one department are visible immediately to others. The capabilities provided by Concurrent As-sembly Mock-Up and CADDSControl allow multiple designers to access

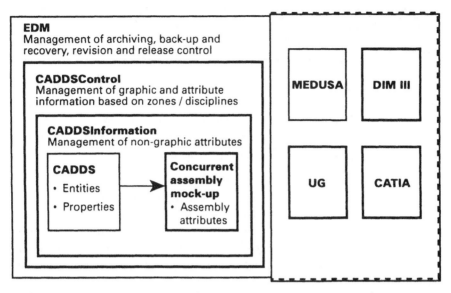

Fig. 10.3   Computervision assembly information management products.

other parts of the design or assembly so that they have the most up-to-date information. While Concurrent Assembly Mock-Up supports assembly information, CADDSControl provides tools for the management of models, drawings and associated geometric zone and functional relationships.

CADDS also leverages data capture in the engineering and design environments, and provides increased integration with other elements of the product process. CADDSInformation and CADDSControl work together to manage a large volume of non-graphic, graphic and attribute data, using standard RDBMS look-up tables. This allows sharing of data between CADDS applications, as well as with other corporate databases such as MRP.

Using the capabilities provided by CADDS, organizations can develop a parts catalog that supports queries for similar designs or investigation of the impact of engineering change orders. These products also support the design process through management of assembly information, supporting multiple versions of the same item in work, and by tracking the design process from initiation through release of the product. CADDS also supports the product specification process through its control of access and management of assembly information across the entire organization. The following pages provide an overview of these products.

10.5.1   Attribute-level management

CADDSInformation is an attribute management system that provides tools for the storage, maintenance and retrieval of attribute data

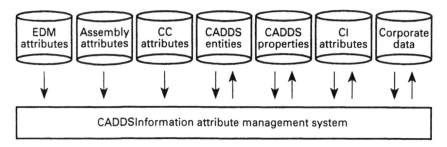

Fig. 10.4   CADDSInformation attribute management system.

associated with CADDS models and drawings. This software supports the parts catalog by providing the ability to extract information on parts and features, as well as the ability to perform where-used queries to determine the impact of changes on products. In addition, CADDSInformation supports the product specification process by making detailed design information available when required.

CADDSInformation manages a large volume of data using standard relational database management system (RDBMS) look-up tables. The Structured Query Language (SQL) is used to retrieve data from these relational tables. Because CADDSInformation relays data between the CADDS data base and the RDBMS, there is no longer any need for organizations to duplicate information inside and outside of CADDS – such as for a corporate MRP system.

One of the advantages provided by CADDSInformation is that access to attribute data is provided through data retrieval from within CADDS or by using the RDBMS. This means that CADDS users are not required to understand RDBMS technology or SQL, and the RDBMS user is not required to learn CADDS.

CADDSInformation provides transparent storage and retrieval of attribute data, including the capability to generate reports from any combination of attributes from multiple database sources. In addition, users can highlight, annotate and update graphic model and drawing data based on attribute data search criteria. Figure 10.4 illustrates the information flow between the CADDSInformation attribute management system and other databases.

Each part and entity linked to CADDSInformation is assigned a unique identifier. The unique part and entity identifier combination is used to link graphic data to related data stored in other databases – CADDSInformation maintains a set of RDBMS tables, called the associated database, for storing these links. CADDS properties can be stored in the CADDS graphic database in the traditional manner, or attributes can be stored in a predefined RDBMS table in the associated database. Attributes can also be stored in corporate RDBMS tables using a user-defined shell script, which can be executed by CADDSInformation commands.

One of the major benefits to be gained from CADDSInformation is maintenance of data integrity. CADDSInformation synchronizes the data stored in the graphic database with non-graphic attribute data stored in the associated database through a set of relational tables that maintain data links between them.

10.5.2   Assembly-level management

Concurrent assembly mock-up is an extremely effective tool for developing electronic mock-ups of large, complex assemblies. Issues of configuration, fit and function can be addressed and resolved by project design team members working in concert. Concurrent assembly mock-up software creates multiple views of the assembly or product structure, which can be stored as part of the parts catalog. In addition, this software supports the product design process by managing assembly information so that variations of the same item can be created. Concurrent assembly mock-up's product structure also allows designers to easily record, track and manage attribute information such as bills of material or labor-tracking data, which can be stored directly in the product structure.

Concurrent assembly mock-up creates a unique database to store all information related to an assembly. The assembly database includes hierarchical relationships between the various elements of the assembly, including their assembly-specific names. In addition, concurrent assembly mock-up provides a graphical representation, or product structure, of the assembly's sub-assemblies and components, and their interrelationships. This logical diagram of the overall assembly is called the assembly navigator. All users can view the assembly navigator structure concurrently and use it to lock, activate and ask questions about the assembly and its constituent parts.

File names of CADDS part models are associated with their corresponding components in the product structure, as well as the positions and orientations of each component in the part model of the assembly. The visibility or display characteristics of the part model are also contained in the database. In addition, users may define and associate attributes with assembly components, which can then be managed with CADDSInformation.

The assembly navigator displays the hierarchy of the assembly, clearly showing how parts relate to each other. The assembly database contains the name and number of occurrences for each part model, the orientation of each instance and any assigned attributes. Concurrent assembly mock-up's easy-to-use features allow the team to build a product structure with any level of complexity. New assemblies can be created, or an existing assembly can be edited, by adding or editing nodes in the assembly navigator. Modifications to complex assemblies may be structured to affect several component instances or only a single, selected component.

In addition, the graphical representation of the product structure allows any member of the design team to access part models associated with components in the assembly. Users, for example, can display geometry and query information from the model, and other functions. Similarly, users can activate a part model for design detailing or modification. Anyone activating a model temporarily 'locks' it, preventing other members of the design team from modifying it at the same time. All other members of the team continue to have full read access to all activated models, and can 'refresh' their assembly display to see what changes were made once the locked part is filed back into the database.

Concurrent assembly mock-up allows the user to simultaneously view both the product structure and the geometry of assembly parts in separate windows. The graphics of one window are linked to their counterparts in the other window. This linkage simplifies and streamlines the diverse range of activities that must take place in order to develop an overall product design process.

Concurrent assembly mock-up also allows users to define exploded views of the assembly geometry in order to position the part model geometry to meet specific design requirements; the user may then switch back and forth between the exploded assembly view and the assembled configuration. Any number of named, exploded assembly views may be created. Exploded assembly views serve a wide variety of functions. The designer may wish to create a view in which one or more parts are shown outside the assembly. This is a useful method to isolate parts of detailing, then switching to the standard view to check fit and configuration. Other uses of this feature include creation of exploded views to examine assembly procedures, display of sub-systems in different phases of operation and display of an assembly from different vantage points.

10.5.3   Zone/discipline-level management

CADDSControl allows designers to define zones for an overall project and associate discipline information and access privileges with users and individual parts or designs. It then provides techniques and changes management facilities for visualizing and controlling subsets of the overall information on the basis of those zones, disciplines and access rights. For example, it allows a user to request information on all changes to structural elements within a particular zone.

Working with large numbers of models and drawings from many engineering disciplines can be awkward in a concurrent engineering environment. For this reason, engineers require tools to integrate multiple-discipline designs and manage their change. CADDSControl partitions a large project geometrically into more manageable zones and allows model data to be classified by function, or discipline. In addition, CADDSControl provides easy retrieval of model and drawing data based

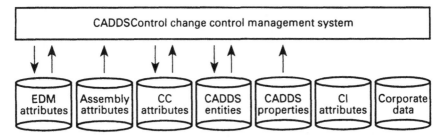

Fig. 10.5 CADDSControl data management.

on zone and function criteria, as well as ensuring project data integrity by controlling user access to specific zones and functions, see fig 10.5. Users have access to both released project data and the latest design changes, since CADDSControl highlights model changes graphically or in report form based on changes propagated by others.

Versions of CADDS graphic databases are transparently stored in EDM by CADDSControl; in-work versions are maintained by the end user in the traditional EDM manner. When all component models are linked and posted to the project database, CADDSControl can also filter the display of assemblies based on zone and function criteria. CADDSControl uses a set of RDBMS tables, called the project database, to track objects based on zones and functional classifications.

CADDSControl constructs objects from graphic entities in the CADDS graphic database based on object construction rules. CADDS entity relationships and properties can be used to define rules to construct these objects, which are tracked by saving versions of the graphic databases, called snapshots. Each object posted to the project database is classified by function, sometimes called a discipline. The volumetric extents of each object are also stored for zone inclusion/exclusion checking. CADDSControl maintains versions of objects in the project database for change-control purposes.

Each user links to CADDSControl for access to a particular set of zones and functional classifications, and can choose to view either released or in-work versions of the data. CADDSControl determines which objects are to be displayed based on zone inclusion/exclusion checking and functional classification criteria, and finds the correct versions of the CADDS graphic databases associated with each object.

CADDSControl also detects the entity changes between two versions of a CADDS graphics database and records changes each time objects are posted to the project database, so that other users can be notified. At that time, the CADDS graphics database is also time stamped and programmatically filed for synchronization purposes. A version of the CADDS database, called a snapshot, is created at the time of posting and stored both locally and in the EDMVault.

Although CADDSControl project data can also be used to generate project-wide reports, it is best to use CADDSInformation for this purpose, as well as for generating reports based on other RDBMS tables, such as corporate data tables.

10.5.4   File-level management

EDM is a tool for sharing engineering data throughout an organization under access, management and integrity control. It provides mechanisms for access and security, release and revision management, maintenance of data integrity, and intergroup communications through notification and project process automation. It also provides a stable and powerful network-wide integration platform that allows customer and application access to data through a client/server architecture.

EDM is designed to support multiple users in workgroups or departments, and is scalable to provide enterprise-level support. This software supports large amounts of on-line data, providing data sharing and co-ordination. For this reason, EDM can be used as the repository for an organization's parts catalog. In addition, EDM tracks and controls multiple, in-work activities in engineering organizations and provides cross-departmental co-ordination of activities.

EDMVault allows users to relate a group of files to each other using a feature called **file sets**. If file sets are defined carefully, the user can easily determine which other files are affected when a change occurs in one of the file set members. A file can be a member of more than one file set, and a file set can be nested within another file set.

EDM also supports user-definable attributes to give users the ability to associate user-defined data to EDMVault objects. An object is any piece of information under the control of EDMVault (i.e. part, file and/or file set). Attributes are important because they describe the **contents** of a file or file set. Attributes can also be used to convey other important information, such as manufacturing specifications or information necessary for a bill-of-materials report (e.g. part or serial numbers). The EDM attribute management sub-system consists of the following components.

- At the lowest level is the **attribute**. An attribute is similar to a column of a table in a relational database. It has a name that is used to refer to its contents and a data type that determines what values it can represent.
- At the next higher level is the **set**. A set is similar to a table in a relational database. It contains attributes and sets that for one reason or another must be grouped together. Attributes as set members have a default value and can be required or optional. If an attribute is required, the EDMVault user must assign a value or the association step will fail.
- At the highest level is the **rule**. Rules are used to determine what user-defined attributes can be associated with an object.

The attribute management rules processor is an exceptionally valuable tool. A project manager or system administrator can define a 'prompt-and-respond' session that must take place whenever a file or file set is created. Additional attributes can be added at a later time, if required. By using rules to establish a fundamental description of a file, information for bills of materials or 'where-used' reports can be easily obtained using EDM's interactive query facility.

EDM maintains two databases: a product database containing product design data and a control database containing the information needed to manage the product data. Both the physical file (the product database) and the metadata (the control database) are kept synchronized at all times, so that consistency is maintained between the files and their control information.

EDM provides significant benefits due to its unique method of storing files. Any kind of file can be stored in EDM, including drawings, bills of materials and custom reports. Data is stored within the EDM vault as individual files and as groups of files or file sets.

10.5.5 Configuration management

A new member of the EDM product family currently in development, called EDMProductManager (EDMPM), specifically addresses the areas of product structure, manufacturing configuration management and product life-cycle management – including the creation and management of bills of materials, parts catalogs and other ways of viewing the product structure. This product can be thought of as the catalog of all parts and assemblies that are used in the production of a company's products, and the specification of how these parts or assemblies are used together (product structure) to produce buildable configurations (configuration management). In addition, the database holds information that controls access to the product specification data and its management (project control, review cycle, change control, etc.) through the development process (product life cycle).

EDMPM is used in conjunction with EDMVault, which stores the bulk data related to parts and assemblies, such as sketches, engineering drawings and CADDS part files. EDMPM is intended to be a high-end bill-of-materials and configuration management database that will solve the problems of relating a particular version of an assembly to another relative to a particular product configuration.

10.6 CONCLUSION

Customers of support tools for concurrent engineering are seeking ever-higher levels of automation and integration with systems that help automate increasing percentages of their design, development, test and manufacturing processes. The combination of tools described in this chapter will enable customers to electronically mimic the processes in place today, without significantly impacting current operations.

Working together, CADDSInformation, concurrent assembly mock-up, CADDSControl and EDM software allows integration of conceptual design, product specification, engineering, detailing, analysis and test, manufacturing, bill-of-materials, MRP, engineering change order, configuration management and administrative activities around assembly or product structure information. The applications that support each of these areas can be integrated with a common information management system, which is open for integration with other systems in the environment.

CHAPTER 11

The role of knowledge-based engineering systems in concurrent engineering

J D A Anderson

The world market is clearly changing The outdated view of 'do you want it fast or do you want is right?' has been replaced by a demand for increased customization, faster delivery and higher quality products without losing control of costs

A significant level of effort is therefore being put into the concurrent engineering process There is a focus on looking at the time spent on the activities involved, seeing how that time can be reduced and improving quality To put it more simply achieving more with fewer people in less time

This is where knowledge-based engineering (KBE) can make a significant contribution

11 1   WHAT IS THE ROLE OF KBE SYSTEMS?

To achieve more with fewer people, significant impact can be made by first understanding the engineering process and secondly by refining it from experience However, it is often time consuming to gain this understanding The problem is made worse because people are too busy carrying out real designs to concentrate on the process of how they are actually doing each design The time spent away from actually designing can perhaps be justified only if we can make sufficient impact upon the future time scales and the quality of the design This is the role of knowledge based engineering systems

A KBE system can be used to store the process used to engineer a product This is called a product model Once this has been done, the product model can be used to carry out much of the engineering automatically (see Figure 11 1)

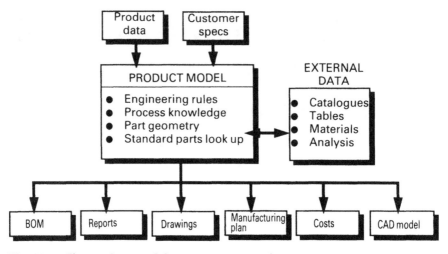

Fig. 11.1 The product model.

11.2 WHAT IS A KNOWLEDGE-BASED ENGINEERING SYSTEM?

The best analogy for a KBE system is to think of it like a new engineer joining the department. Usually, the first tasks assigned to him or her are those which can be learnt in a short space of time and will be carried out sufficiently often to justify the effort. Selecting suitable KBE system projects uses exactly the same criteria. We need to balance the time taken to teach a KBE system how to do something against the benefits of it being able to carry out the task very quickly, consistently and to best practice.

Another way of looking at this is through considering how a typical engineer spends his or her time. The majority of time is spent doing routine engineering, following an engineering process which is already familiar and carried out as a matter of routine. Traditionally, this often leaves little time for creative engineering, i.e. thinking of new and better ways of designing the product.

Let's take a simple illustration. Consider the design of a load-bearing component. It may occur to the engineer that changing the wall thickness of the component may give a lighter design. However, this creative idea may take several weeks to investigate. The product has to be re-designed, including increasing the stiffening to provide sufficient strength, rigidity and life of the product. Calculations like this, plus basic considerations of manufacturability, can absorb a major part of the available time.

A product model is designed specifically to automate such 'routine' tasks. KBE is not trying to automate the creative part of engineering. It is, in fact, a strongly held belief that the engineer should be creative rather than the computer. A KBE system is therefore not trying to be artificially intelligent (AI). Neither should it be confused with an expert system, which represents a significantly different technology for a different purpose.

Instead, a KBE system is focused purely on the automation of the routine engineering area, letting the engineer be truly creative. This is exactly what good engineers want. Now, with our previous example, the engineer is free to decide on a 'what-if' by changing the wall thickness. The product model can be set up to re-engineer and re-design the product, following the process previously described. Our engineer will then be presented with the new design, costs and weights in a matter of minutes instead of weeks. He or she is thus able to do perhaps another 20 to 30 'what-ifs' in the same afternoon.

The product model is therefore used primarily to capture specific design and engineering processes. This may cover all the areas that we would describe to the new engineer in the previous example, if we wanted him or her to carry out the same task. This would typically include:

1. design constraints;
2. any calculations that need to be performed;
3. how and where to use data and standard parts from catalogues;
4. how and when to use external programs and databases;
5. manufacturing constraints and any other downstream requirements.

Some of this knowledge would relate directly to geometry but a significant amount would be non-geometric. All this knowledge is stored in the product model.

11.3   WHAT RESULTS CAN BE EXPECTED FROM A PRODUCT MODEL?

If we look again at Figure 11.1, we can see that the product model can contain the engineering process, any downstream requirements and, of course, the process to create the geometry of the design. In addition, we often want to build in the knowledge of when and how to select information and standard parts from catalogues and tables.

Like the tooling example, we often start the process of designing something by being given a specification that can include both geometric and non-geometric constraints and requirements. Both forms of input can be passed to the product model.

Outputs or results from the product model can be virtually whatever is required (see Figure 11.2). The drawback however, is that for each output a varying amount of time has to be spent defining how to generate it in the product model. This, again, is exactly analogous to our new engineer who needs an explanation in order to generate the results we want.

Thickness = f(depth × pressure/max deflection)

NC tool path

Query parts catalogues

Share CAD geometry

BOMs costs reports

Engineering drawings

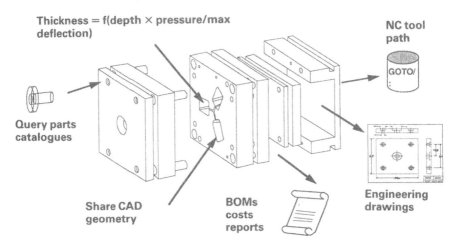

Fig. 11.2 Capturing engineering knowledge

Some output is easy to generate since the product model has already worked out the information required; a bill of materials is an obvious example, generated with the geometry that has been used to do the engineering. Cost reports, however, will require the extra time it takes to instruct the product model how to cost the design to the required accuracy. Some companies have built product models to generate manufacturing plans and inputs to other programs, as well as update records for external databases and produce a wide variety of engineering reports.

11.4   CAN KBE BE USED TO OPTIMIZE A DESIGN?

KBE can be used as a tool to enable the engineer to optimize the design in two ways. Firstly, the engineer can easily change any of the default values or inputs that drive the product model in order to search for a better design. A well-designed product model will provide facilities for the end-user to modify any design aspect that needs changing in the search for the best overall design. The product model then automatically re-designs the part, ensuring all the engineering and other considerations are met. It can then present the end-user engineer with critical information, such as weight and cost, that might be used to measure the effectiveness of his or her idea. Clearly the KBE system will allow many creative 'what-ifs' to be done in a relatively short space of time.

The second area relates to the product model itself. Unlike people, computers are good at doing repetitive things very fast. This ability is invaluable in the search for optimized solutions. For example, let us assume there is a key dimension for the design, perhaps representing the position of a key component. Conventionally we may need to guess the best position of a component using our prior experience in order to continue with the design.

With a KBE product model, however, we simply need to say that we know the best position lies within a particular range of dimensions. The computer can then search for the best value. This means that the product model might sample several hundred values. For each value it can generate a full design, along with the weight and perhaps even the cost of that particular design. It can then automatically present us with the optimum design for that one dimension value in terms of weight or cost. Sometimes the searching involves varying a number of values or selections at the same time in order to find the best design.

Fortunately, the KBE system is clever enough not to completely re-design and re-engineer the part for each iteration, but only re-work the areas that are different from previous iterations. This significantly reduces the load on the computer, which allows many more iterations to be considered in a reasonable amount of time.

11.5   WHAT TECHNOLOGY IS USED TO REDUCE THE TIME NEEDED TO BUILD A PRODUCT MODEL?

Clearly one of the main drawbacks of using a KBE system is the time it takes to build the product model. There are, however, a number of important aspects to the system that allow the actual building time to be far less than might at first be imagined. One of the key technologies underpinning KBE systems is the object-orientated approach (see Figure 11.3). This allows a word to be defined that has real meaning for the engineer, for example a 'pressed part'. To return to our earlier analogy,

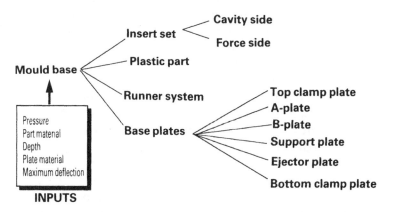

Fig. 11.3   Example product structure.

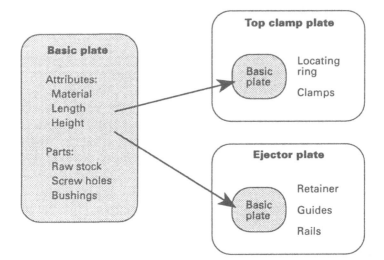

Fig. 11.4   Using 'kind of' basic plate for both the top clamp and ejector plate.

we might say to the new engineer that a pressed part must have a draw angle maintained for all faces of the part. In other words all side faces must be designed so that the part can be pressed and extracted from the tool.

When we talk to a real person, once we have explained the considerations for a pressed part we do not expect to have to explain them again. In fact, with any future designs we only expect to say that it is a 'kind of' pressed part for all our previously described considerations to apply.

It is exactly the same with a KBE system. Once we have described what we mean by a certain word, such as 'pressed part', it can be used to describe any design simply by saying that the design is a 'kind of' 'pressed part' object. This 'kind of' behaviour is extremely useful when building up a product model since any object – once explained – can be used anywhere. Other examples of this object-orientated approach are engineering calculations such as a 'kind of' Euler buckling calculations, or a 'kind of' 'basic plate' which might imply a number of design constraints and manufacturing requirements (see Figure 11.4).

The same approach is useful for actual features and parts where they might be described as a 'kind of' web or a 'kind of' rivet.

Apart from this 'kind of' description, features and parts can also be described as 'part of' something else. For example, a web is part of a stiffener, which is part of a wing-box, which is part of an aircraft. This reflects a normal engineering viewpoint. When a grow-out is described as part of a stiffener, an engineer will immediately assume that a number of characteristics of the grow-out are the same or inherited from the parent object stiffener. The material is a good example.

In everyday language, an object is often described as a 'part' of something; it is also sometimes a 'kind' of something – with, perhaps, some unique differences specified. KBE systems work in exactly the same way, in that both 'kind of' and 'part of' descriptions are used to drastically reduce the amount of time required to build product models. In addition, many additional powerful software strategies have been adopted to further reduce the time and skill required to build a product model. Sophisticated product models have been shown to be over ten times faster to develop than an approach using conventional programming techniques. In addition, it is very much easier and quicker to maintain them afterwards. It would take too long to detail these advanced software features here, but it is worth pointing out one of the major benefits. Unlike conventional programming, detailed specification of the engineering process is not needed before starting to build the product model.

11.6 SHOULD THE ENGINEERING PROCESS BE DEFINED IN DETAIL BEFORE STARTING TO BUILD THE PRODUCT MODEL?

It is not a good idea to spend weeks, months or even years trying to establish the best practice design process before using a KBE system. The reason is simple: it is fundamentally difficult for people to try to describe exactly how they do anything. This is why conventional program specifications nearly always have to be modified once the carefully structured program has been shown to end-users. Typically, several iterations later the program starts to do what the end-user wants, but the program structure has become increasingly messy in the interim. There are also ensuing difficulties in debugging and maintaining the program.

Fortunately KBE systems were designed to be used by engineers who want to start defining a process straight away, look at the resulting design, criticize it and then change it until they are happy with it. In other words, to develop the product model by successive refinement. This allows the engineer to criticize the early designs from the product model, using these as a basis for refining and describing the process. People find this far easier than trying to get something right first time. The expert's ability to criticize designs is, after all, well known!

The approach of successive refinement in building the product model is particularly powerful when it comes to enhancing of the process later. This becomes a simple further refinement of the product model. It should be remembered that although a product model may be released for design use, product models will never be complete. They should continue to evolve in order to represent the latest engineering ideas and processes.

11.7   HOW CAN THE KBE SYSTEM BE INTEGRATED INTO AN EXISTING COMPUTING ENVIRONMENT?

Typically any engineering department will already have a number of computer programs and databases that they use in conjunction with a CAD/CAM system. Clearly any new engineer in the department would be expected to use these existing tools rather than create their own.

Similarly, the KBE product model should be able to use these existing systems as part of the engineering process. The problem facing the KBE system vendors is that every customer wants the product model integrated with different programs and databases as well as different CAD/CAM systems. A good solution to this problem is the creation of integration tool kits. This concept drastically reduces the time it takes to build a specific integration.

These tool kits include one for accepting geometry inputs from an external source, one for calling external programs and databases and one for creating output, in any form, for both geometric and non-geometric results. These same tool kits are also the basis of specific interfaces for popular CAD/CAM systems, external databases such as Oracle and external programs such as Parasolid. In addition, the same tool kits are made available to the end-user as standard products. This allows specific integration to be achieved relatively easily by the end-user company.

It is also important that the product model should be able to be run from a variety of terminals and workstations, using the user interface from other systems. This is achieved using another tool kit which allows the product model to be run programmatically over a network from another system. Again, this tool kit has been used by the KBE system vendor to provide the interface software that allows the KBE system to be run directly from within a CATIA or Unigraphics CAD system environment.

11.8   WHAT IS THE DIFFERENCE BETWEEN KNOWLEDGE-BASED ENGINEERING AND PARAMETRIC GEOMETRIC MODELLING?

Let us first consider whether the geometric descriptions of a particular product actually represents the design or is really one of the results of the design process. For many components and assemblies, the final geometry is largely the shape or configuration that meets all the engineering and manufacturing considerations. The final geometry can therefore often be considered to be the result of the design process. This is a very useful result, as it is the basic communication vehicle by which the rest of the company knows what has been designed. In addition, the geometry is often used by many downstream applications involving manufacturing, production, technical publications, purchasing and so on. The problem occurs when we need to change the design. Let's consider how KBE systems can streamline a re-design so that all the engineering and manufacturing considerations are still met.

Assuming that these considerations have been captured in a product model then all we need to do is simply change the inputs – geometric or non-geometric, run the product model and have new geometry automatically created. It will conform to all the original considerations. This is effectively knowledge-based engineering.

Parametric geometric modelling, on the other hand, means that we can change the geometry of the design very quickly. If we do so however we no longer have the assurance of knowing if all or, indeed, any of the original engineering and manufacturing considerations have been met (see Figures 11.5 and 11.6). Parametric geometric modelling is a powerful tool when the engineering (and other requirements) are so well understood that the engineer knows exactly what geometric modification needs to be made in order to meet the engineering requirements and achieve the best compromise.

A good example that illustrates the essential difference between the two processes is the design of tooling. A particular type of tooling is normally designed using the same engineering process. Each tool however can be geometrically quite different from another tool. Clearly, this is because different tools are for different parts.

The geometry of the part is therefore one of the main inputs to the tooling process. Other inputs might be the production volume and the limitations of the machine that is to be used. The geometry of the tool design results from taking this and other inputs, following the tooling design process and generating the tool design geometry. When the tool design process is described in a product model, it can be used to

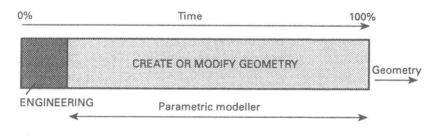

Ideal:
  • when the engineering content is easily understood

Fig. 11.5 Expanding of parametric modelling.

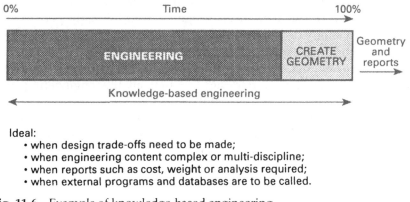

Ideal:
- when design trade-offs need to be made;
- when engineering content complex or multi-discipline;
- when reports such as cost, weight or analysis required;
- when external programs and databases are to be called.

Fig. 11.6   Example of knowledge-based engineering.

automatically design the tools for all the parts covered by the process. The idea of building a parametric geometric model of one tool and trying to use this to design another tool for another part that will be made on a different machine is clearly not sensible, particularly as the subsequent tools may have totally different topologies from the first tool, i.e. they may not even look similar (Figure 11.7).

Fig. 11.7   What is good knowledge-based engineering?

11.9   WHAT IS AN APPROPRIATE KBE SYSTEM PROJECT?

Simply put, an appropriate KBE system project is one where the costs of the project are significantly outweighed by the benefits resulting in a good return on investment (ROI) (see Table 11.1). What are the costs? Clearly, the obvious ones are the KBE system software, consultancy, training, computer hardware and maintenance. The major cost, however, is the cost of personnel to carry out the project. It is therefore very important to obtain skilled advice on how long projects are likely to take before starting on them. It is also worth pointing out that it is important to choose the right people to carry out KBE system projects. Different people doing the same project can take half or even a third of the time. The right choice can result in major cost savings.

Table 11.1 Benefits versus costs

Benefits	Costs
• Manpower saving	• KBE system
• Shorter lead times	• Manpower
• Higher quality design	• Training
• Disseminate expertise	• Consultancy
• Understand the process	

So, what are the benefits that can be achieved for an appropriate project? First, the process captured will be carried out in a tiny fraction of the time it takes manually. The value of this is a function of how much time is saved per use of the product model, multiplied by how many times the product model will be used. This can result in significant personnel savings, perhaps allowing more people to be allocated to building future product models – with even greater benefits. Another benefit is that the results of the product model are available for use much sooner than before. This may have a significant effect on the overall lead time of the design program or response to a tender inquiry.

The second major benefit to most users is that applying KBE to a particular project means that at the end the process involved is much better understood. This process of establishing best practice can be a major benefit by itself.

Once they have understood the process and established best practice, many companies now find that it is beneficial to be able to distribute this process in the form of the product model. This may involve passing it to affiliated companies or to other departments in the company. For example, a bid proposal system might even be made available to the sales department to generate customized designs. Alternatively, it may simply mean that new engineers and other less skilled people can use the product model to the same level as the process captured. But perhaps the biggest benefit to most users of the technology is that they can achieve a higher quality design. Below is a list of some of the ways in which these quality improvements can be achieved.

- An optimized design in terms of, say, cost, weight or performance.
- Consistent designs from different designers, resulting in reduced variations and attendant manufacturing and service benefits.
- Less re-work and fewer modifications required due to following best practice, as well as ensuring downstream requirements are met, e.g. manufacturability and serviceability.
- Design freedom kept open much longer in the design cycle than possible with normal design.
- New design ideas, materials, etc. can be incorporated and used widely.

• Rush work does not mean some of the design considerations are forgotten.

11.10 WHAT BENEFITS HAVE BEEN ACHIEVED BY EXISTING USERS USING A KBE SYSTEM?

The following projects show examples of some of the benefits described.

Company	Benefit example
Lotus Engineering	Keeping design freedom open much longer
Jaguar	Design for manufacturability
Lucas Aerospace	Optimized design

Lotus Engineering recently undertook the design of the bicycle used by Chris Boardman to win his gold medal for Great Britain at the Barcelona Olympics. This design involved a new approach centred on a monocoque frame and include use of sophisticated CAE techniques for stress analysis and aerodynamics. The problem was that when the design engineers started the design, they did not wish to fix the key dimensions of the frame since they did not know the best values to choose.

Instead, they decided to capture the whole design process using ICAD (intelligent computer-aided design system, marketed by ICAD Engineering Automation Ltd.), starting from a blank screen. In the same period in which they might have expected to design, analyse, build and test one bike using CAD/CAM, Lotus was able to successively design, analyse, build and test four bikes. On the last iteration, they were able to make fundamental design changes less than four weeks before the Olympics.

The use of ICAD to capture the process of the design, links it to analysis and ensure manufacturability, allowed Lotus to keep major design decisions free 80% of the way into the project. Conventionally, these decisions would have to have been fixed at the beginning of the project when insufficient information was known to make the best choices.

The Jaguar example involves the process of designing the stiffener for a body panel, such as a car bonnet. The inputs to the process are the styled geometry of the exterior panel along with inputs from the engineer about the position and depth of the stiffeners, the hinge positions and requirements for the closing of the bonnet. The surface panel shown takes about two hours to be produced automatically by the product model and would conventionally take around two months on the interactive CAD/CAM system.

Conventionally, once the panel has been produced it is shown to manufacturing feasibility in order for them to establish its suitability for pressing. Clearly any changes may then require further weeks or even months of interactive CAD/CAM work. One of the requirements, therefore, built into the product model, is for all the surfaces of the stiffener to be designed with the appropriate draught angle relative to the specified press tool direction. This will help ensure that the part can be extracted from a press tool. The important point here is that although the design engineer has control of the design, the product model will only create the geometry that conforms to the manufacturing requirements built into the system.

The Lotus and Jaguar applications illustrate the benefit of including multi-disciplinary requirements in the product model. Some other users have taken this further by including stress, thermal vibration and other types of analysis into the product model. Again, the benefit is that the geometric result created by the product model will already conform to these various requirements. This can give a huge saving in lead time when, conventionally, many weeks or months can be lost while the various departments involved slowly iterate the design so that it meets everybody's requirements. It should be recognized that – providing the product model is capable of generating an automatic design as good as that normally achieved half-way through the conventional design process – the total lead time of the design will be halved. In other words, one does not have to wait until the product model can produce the final design before achieving significant benefits. The last example is from the actuation division of Lucas Aerospace and involves the design of a gearbox used as part of the control system for the flaps and slats of an aircraft wing. An interesting aspect of this design is that the process built into the product model includes a number of levels of iteration in order to establish the best design.

The effect of all these iterations is to automatically create around 400 designs before the final design is selected. This includes running the in-house analysis programs on each design automatically. All of this is transparent to the end-user engineer who simply wants to generate the appropriate input specification and then examine the final design presented.

A wider selection of KBE application examples is shown in Table 11.2. These applications are either being developed now by industrial companies or have already been developed by them.

11.11 WHAT IS THE FUTURE FOR KNOWLEDGE-BASED ENGINEERING SYSTEMS?

Perhaps the more important question concerns the future for those companies who choose to exploit KBE technology wisely, given that they

Table 11.2   Examples of knowledge-based engineering applications

	Bid proposals	Component and assembly design	Generative tooling and planning
Aerospace and defence	Passenger compartment layout	Aircraft turbine blades and components Satellite antennae	Composite tooling Turbine compressor tooling Sheet metal tooling
Automotive	Vehicle packaging	Body panels Cylinder heads Crankshafts and connecting rods manifolds	Stamping die layout Forging Plastic tooling layout
Consumer products	Replacement windows HVAC control systems	Medical prosthesis CRT glass Vending machines	CRT glass moulds Plastic mould bases HVAC control software
Industrial equipment	Power distribution equipment Plant layout Electric motors Industrial machines	Heat exchangers Fans Railway bogies Drilling tools	Bending mandrel N/C Assembly drawings Panel breakdown

have already achieved significant benefits. The wider use of the technology should allow these companies to spend more time on understanding and refining the process by which they design and manufacture products. This carries obvious competitive advantages. In addition, less time is spent on carrying out actual designs, allowing costs to be reduced and quality to improve. This, again, increases the competitive advantage.

As KBE evolves, it will develop ways by which users can build their applications faster as well as use the technology more easily, particularly as end-users of product models. Wider deployment of existing knowledge bases is obviously extremely important in order to achieve the maximum benefits. This wider deployment will often require the KBE system to be hidden behind existing systems with which the potential end-user is already familiar.

An obvious example of this is the CAD/CAM system. A close integration with the CAD/CAM database needs to be matched with the ability for the product model to be driven from within the CAD/CAM system user interface. It is also important that a KBE system complements the conventional interactive CAD/CAM systems in which so many companies have already made significant investments.

This close integration of CAD/CAM systems can only be achieved with the co-operation of both vendors. I believe the future will therefore see KBE systems even more closely integrated with selected CAD/CAM systems to provide a next generation design environment. Unfortunately, not all CAD/CAM vendors have an open systems approach (regardless of the claims from their marketing departments!) which means that they do not provide the necessary facilities and support to achieve integration with third-party systems These vendors will then be forced either to develop their own KBE system functionality or to leave their customers with IGES type translators to third-party software

As some CAD/CAM vendors will be unwilling or unable either to provide an open system or to develop their own KBE system, I believe the future will see two types of CAD/CAM systems At one level will be those vendors with an integrated KBE system, the second level will be those without Any users of a second-level system may find themselves at a significant competitive disadvantage in the future

11 12   CONCLUSIONS

In an increasingly competitive world, many companies need to achieve more with fewer people while reducing costs and improving quality, the very essence of concurrent engineering The wise use of KBE system technology provides a platform to help achieve these objectives Today KBE systems represent a technology that allows people to achieve more in a shorter time, while providing a higher quality design with the possibilities of better cost control Companies who increasingly use this technology for appropriate applications should therefore gain an increasing advantage over any competitors not exploiting the technology to the same depth

FURTHER READING

1 Stalk, Time (1988) The Next Source of Competitive Advantage *Harvard Business Review,* **July–August,** 41–51
2 Tortolano, F W (1991) Boeing's Big Twin *Design News* **26 August**
3 Anderson, J D A and Davies, P F (1991) Knowledge-Based Engineering Systems in Practice *IMechE,* **C429/043**
4 Cinquegrana, D (1990) Intelligent CAD Automates Mold Design *Mechanical Engineering Magazine* **July**
5 CAD/CAM/CAE/GIS Industry Services Staff (1992) Market Analysis *CAD/ CAM/CAE/GIS Forecast Update Dataquest* **October 19,** CCAM-COR-DP-9206
6 Gregory, Annie (1992) Separating Fact from Fiction *Manufacturing Breakthrough* **November/December.**

CHAPTER 12

Software solutions for concurrent engineering II

S. Schedler

12.1 INTRODUCTION

The concept of concurrent engineering may not be new in itself. After all, it reflects sound engineering practice. What is new, however, is the fact that computer-based tools are now becoming available to turn the concept into reality. As a developer and supplier of these tools, Intergraph is working increasingly today with companies to implement systems which will bring back the level of inter-departmental communications crucial to concurrent engineering, and which has been lost over the years as company structures have grown larger and more complex and systems have been purchased to satisfy the needs – or self interests – of individual departments, rather than the whole engineering workflow.

Put simply, the adoption of concurrent engineering practices means that the design of a product and the systems to manufacture, service, support and ultimately dispose of it are considered together. Clearly, concurrent engineering is linked to concepts such as multi-disciplinary project teams, total quality management and, from a computing perspective, integrated systems built on industry standards, and an open, networked environment.

12.2 DEFINITION

Integration is a word commonly used by commentators on, and users of, information systems. For our purposes it will be useful to focus on a dictionary definition – unification into a whole.

This is a very broad definition. Integration is not just the ability to do things at the same time, and it is not just interfacing one system with another. It is putting things together so that the separate parts work as one. If we start from this definition it already becomes obvious that integration is a powerful tool for improving the efficiency of a business.

12.3   DEFENCE IMPETUS

The acceleration in the development of computer systems to support concurrent engineering and the level of interest being shown in them by manufacturing companies owes much to recent activities within the defence industry.

In the USA, the Department of Defense is engaged in an initiative known as CALS (computer aided acquisition and logistics support). This programme is aimed at 'improving the accuracy, timeliness and use of logistic technical information' by defining standards which will accelerate the use and integration of digital technical information associated with the acquisition, design, manufacture and support of (military) equipment.

Although this initiative is primarily defence – industry driven – the benefits it is designed to bring apply equally in the commercial manufacturing sector, where the standards being developed under CALS for systems to support concurrent engineering are of particular relevance. Indeed, it is the commercial sector which is currently taking up much of what is already available.

12.4   UNLOCKING INFORMATION

This chapter is not technically orientated but it is clear that the first and most basic requirement for unification into the whole is the integration of the systems environment. In other words, we need to rid ourselves of the difficulties presented by islands of automation which have sprung up across companies during the 1980s as small systems were added to provide small solutions before it was possible to implement an overall business systems strategy.

Today, a great many companies have discrete, 'point' solutions for dealing with specific problems, in areas such as mechanical modelling, drafting, engineering analysis, numerical control, electronics design, document publishing and facilities planning and management. They have networks with software and hardware from a variety of vendors and are adding to this data every day both manually, as well as with sophisticated computer-based systems.

The key to unlocking this information and making it available to those who need it, in the controlled manner necessary for a concurrent engineering environment, is based on three main factors:

1. flexible local and wide area networking and file management facilities;
2. application software built on a common data structure or framework;

3. customizable tools for technical information access, and distribution and sign-off procedures.

Ever since the early 1970s Integraph has developed its systems around a common data structure, or file format, which was defined in software known originally as IGDS (Interactive Graphics Design System) and now, in its current manifestation, as MicroStation. The file format definition for this is 'public domain' and therefore freely available from Integraph to independent software developers.

MicroStation forms the nucleus of over 1000 Integraph and third-party developed application software packages, allowing design data created on UNIX workstations, PCs and even minicomputers to be moved freely between engineering disciplines without translation.

Its future is assured. Several recent major contracts commit Integraph to maintain this nucleus file format for the next 30 years – longer than the CAD/CAM industry itself has been in existence!

In addition, Integraph designs its systems to incorporate existing, proposed and *de facto* industry standards – creating an 'open' system environment. Within this open environment, Integraph has implemented standards for operating systems (UNIX, POSIX, DOS), software development tools (X-Window System libraries, OSF/Motif, GKS, PHIGS, ANSI C and Fortran compilers, ADA and others) and networking (Ethernet TCP/IP, DDN, SNA, DECnet, LU6.2, FDDI, GOSIP and Token Ring).

By adhering to these and other standards, Integraph systems co-exist with systems from other vendors which may already be in place. And to permit the exchange of data between these differing systems, Integraph fully supports accepted data exchange standards (IGES, VHDL, EDIF, PDES, CCITT Group 4 etc.). As long ago as 1981, Integraph decided to base its systems on networks, thereby committing to solve the problem of distributed database management. Today, the Integraph network file manager (I/NFM) solves this problem by providing configuration management of a geographically distributed database resident on systems from multiple vendors. I/NFM, working with Integraph's relational interface system (RIS) to support a common access method to multiple databases, provides network file management, privileged access control, workflow management and revision control.

Built on a relational database, Integraph configuration management tools locate and retrieve information regardless of its location on the network. Integraph document scanning, engineering and publishing applications, as well as external sources, feed the database with information in multiple, co-existing formats – raster, planar vector, 3D solid vector, SGML and ASCII.

To satisfy the diverse needs of a concurrent engineering environment, Integraph provides database-independent and network-independent access to a heterogeneous environment through its RIS product. This interface allows simultaneous access to multiple commercially-available relational database management systems, including Ingres, Oracle, Informix and DB2.

For a fully graphic means of database access, Integraph's DBAccess software provides a forms-based, user-definable method for creating intelligent, associative relationships throughout information stored in multiple databases and for extracting data from them. To satisfy the third criteria for concurrent engineering, Integraph's technical information management systems (TIM) support access to data stored on both Integraph and other vendors' equipment. Data can be stored on equipment provided by vendors such as IBM, DEC, Sun or HP. The Integraph TIM environment also allows users on Integraph workstations, VT100 terminals, or other vendor workstations or PCs with VT100 emulation, to take advantage of network-based file management capabilities.

Integraph's TIM systems provide a framework for integration of disciplines, enabling CAD-generated and existing paper-based information to be integrated into the network and providing an important link with corporate libraries which may extend across the world.

12.5 TECHNICAL INFORMATION MANAGEMENT

12.5.1 Introduction

The management and control of technical information is a real and growing issue. As more industrial processes are being automated, more computers are being used and there is a need to distribute data more efficiently. There is a danger that the advantages of automation will be lost as the volume and complexity of the engineering data available and the difficulty of locating and controlling data increases proportionally. As workstations and PCs become available at the workplace and computing becomes more distributed, the logistics of data management become more complex. If this problem of complexity is not addressed, then many of the benefits of distributed processing will be lost.

A technical information management system must manage and control data of varying types, stored on differing platforms and accessed through different communications protocols. It is necessary to integrate the use of technical data within a company, whether the data is CAD-generated, computer-based or on paper, so that it can be used together to ensure that sufficient information is available to support business decisions.

It is important to recognize that information must flow through the workplace, that each process adds to the available data and therefore adds value to the business process. Thus, it is important to be able to

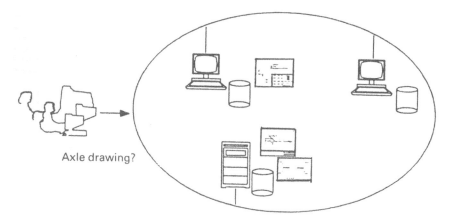

Axle drawing?

Fig. 12.1 Networked workstations and common user access.

integrate data across application boundaries and also to optimize the flow of information between processes in the engineering environment.

12.5.2 Information management requirements

Whereas the introduction of network systems based on workstations has brought many benefits for technical users, it has also created a data management problem. Users are faced with a system which may contain different hardware platforms, operating systems and network protocols, each with its own syntax. So, not only do they have to remember where their data is, but they have to remember what it is called and what they must do to access it (Figure 12.1).

There is thus a requirement for a simple method of accessing data on the network, which does not require a knowledge of syntax or of the location of the data and is based on engineering terminology, such as part number, drawing number, location, and so on. This nucleus file management system should be as simple as possible to implement and should not require any major expenditure on software or consultancy.

In many cases, there is also a requirement for registering data in an archive or repository and providing a method of access control based on standard workflows. This would ensure that data is complete, consistent and protected against unauthorized modification at each stage of its lifecycle. There are also requirements for release control, covering the distribution of data to interested parties, and change control, covering the authorization of revisions (Figure 12.2).

In general, the basic data registration, management and control requirements are similar for most users. Thus, it should be possible to provide a nucleus data repository system which is simple to implement and would not require extensive customization and consultancy.

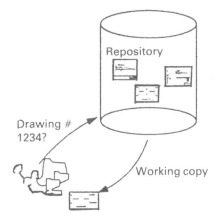

Fig. 12.2   Nucleus data repository.

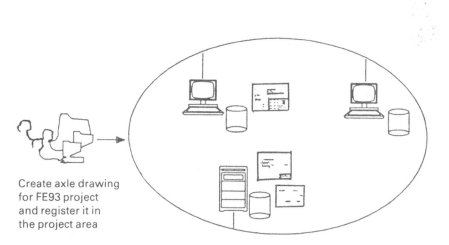

Fig. 12.3   Project group concept.

A certain amount of customization would be inevitable, since differences do exist, particularly in the data input and validation areas.

It is not enough to provide a network data management system, such a system must also be comfortable for the intended users and must integrate into their way of working.

Thus, there is a requirement for a customizable engineering interface to any data management system, to provide the level of comfort and integration required by the users. This would entail a drawing registry system, to provide the basic functionality required in the management and use of engineering drawings (Figure 12.3).

There is also the concept of the 'work group' or 'project group', where several designers or engineers work together as a group in order to complete a project or task (Figure 12.4).

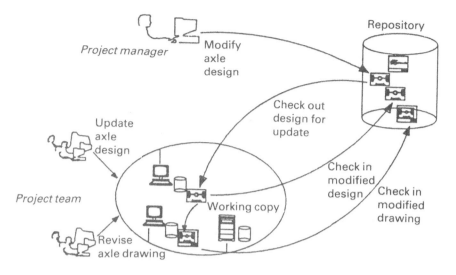

Fig. 12.4 Formal corporate workflow

The group members share data and may share workstations There may be several different types of hardware and software in use within the group The groups tend to work in a fairly informal way, but may be part of a more formal corporate workflow Where application software is used to access data, then the requirements for basic file management, access control, workflows and so on should be tailored for the particular application involved However, it is important that data management software should be compatible across applications Some companies also have a requirement for a corporate data structure, particularly in a multi-discipline environment, which usually requires extensive analysis and customization, but if the nucleus systems outlined in this section were available, then the effort involved would be reduced, since much could be based on the standard offerings (Figure 12 5)

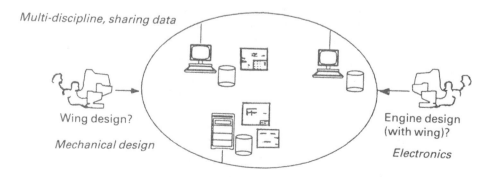

Fig. 12.5 Multi-discipline data sharing

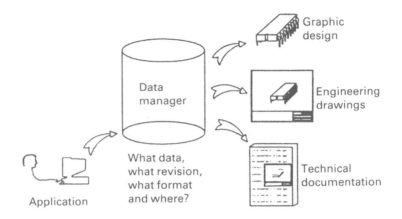

Fig. 12.6   Data manager.

Data exists in many different formats and it is important to be able to find the right data in the right format for the application to be used. The relationships between the different views of the data must also be managed, since it should be possible to indicate where modifications have taken place and what other data is affected by the updates (Figure 12.6).

It is important to be able to use historical data to develop new products, since this can reduce the 'time to market' considerably and help reduce development costs. Developing a new product from an existing similar design is far easier than designing the whole product from scratch and may cut development time considerably. It is also easier to standardize procedures and products if previous designs are used as a basis for new work. Obviously, 're-inventing the wheel', usually in a slightly different form each time, wastes development resources and should be avoided wherever possible (Figure 12.7).

The management of technical information is the key to the successful use of historical data, since it is important to be able to readily identify what data is available and to ensure that all relevant data is retrieved, so that the basis for new designs is as complete as possible.

Where technical documentation data is to be managed, for example in the case of documents consisting of scanned drawings and related text, then the volumes of data tend to be very large, but the processing required tends to be rather simple, such as view or redline (annotate) data.

In a technical documentation system, there is usually a mix of large and small format, CAD data, scanned paper documents, word processing data, text data, data input through translators and so on. Technical documents tend to be complex and many have relationships either with other documents or with their original data sources. These relationships

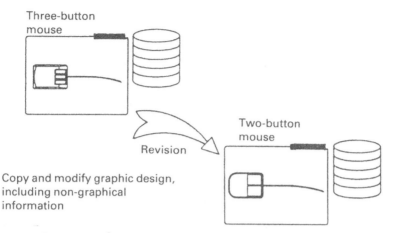

Fig. 12.7 Revision and conversion process

must be maintained, so that comments or maintenance notes can be distributed to all interested parties and documentation can be updated easily when design changes are made (Figure 12 8)

Currently, much of an average company's data is in paper form, rather than in computer readable form A technical information system must be able to manage all types of data, not just that small proportion which is stored in computer files Engineering drawings, documents, reports, etc can be easily converted into a computer usable form by scanning and storing them in a standard format (Figure 12 9)

Technical data of various types should be made available for maintenance and documentation purposes This requirement implies that the

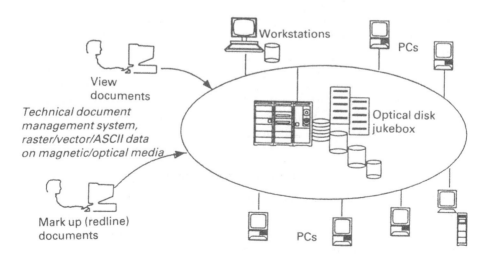

Fig. 12.8 Large and small format data capability

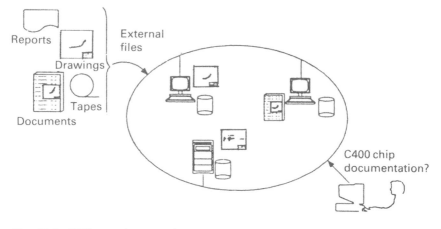

Fig. 12.9 Different data type inputs.

data should be made available in a convenient format for use in technical documentation, since the source document or drawing may contain different information than is required for documentation purposes. However, there should be a linkage between the original data and that included in the documentation, so that modifications and revisions can be reflected and the documentation kept up to date.

If technical documentation is held as hard copies, the bulk involved can require both extensive storage space and a comprehensive indexing system. The maintenance of such documentation can be very expensive and the management of the documentation whilst it is being used is difficult. Documents can be mislaid, misfiled or damaged and if they are copied, then problems with version control and consistency can arise, i.e. which is the latest version and does it show all updates.

In certain cases, there may be logistical problems with the documentation, particularly where the site is extensive. Management facilities can be provided in a computerized technical documentation system and many different types of data can be incorporated. It is even possible to register and index hard copy data in such a system, so that the progression from a manual technical documentation system to a computerized system may be facilitated.

12.5.3   Technical information management requirements

Analysing the user requirements outlined above, it is apparent that there are requirements for several types of technical information management system, which, of course, must be compatible. The major categories of technical information management are as follows.

● A nucleus file manager, so that data files can be created and manipulated in a network without the user having to know where the file is located or what syntax the computer system uses.

- An archive or data registration system, where data files can be saved to secure storage so that access to data can be controlled.
- A modular data registration and management system, reflecting the working practices of the company, where selected data can be presented to either a single engineer or to a project/work group for further manipulation.
- A document management system, so that the many hard copy documents, drawings and manuals can be converted into computer-based forms, and can be used, stored, managed and controlled in the same way as other computer files.
- A fully customizable technical data management system which reflects the structure and working practices of the organization.

Each component should be compatible and customizable. It should be possible to install a standard start-up implementation of each component quickly, so that training, testing and evaluation can begin as soon as possible. It is not intended that such a start-up system should be a solution to all of the technical information management requirements, but that it should be a first step on the path to the solution. It should be possible to customize the technical information management system to suit varied engineering environments and to reflect the most comfortable method of use for the many different types of users of such a system.

The effort involved in creating and maintaining fully customized TIM implementations for individual companies is great compared to that of creating and maintaining basic, customizable packages which can be used by many organizations. Such simple packages can then be used as the basis for fully customized systems. A step-wise implementation plan ensures that the correct solution is found and that the users or their representatives are fully consulted and involved at each step of the process.

12.5.4 Technical information management characteristics

It is possible to extract the characteristics required in each area of technical information management from the user requirements specified above. The characteristics of a nucleus file manager include the following points.

- It should not involve major expenditure, for example, it should not require an RDBMS.
- It should be transparent to the user, i.e. not syntax based.
- It should have a customizable graphical user interface for ease of use.
- It should support VAX systems, workstations and PCs.

- It should be usable from a variety of platforms (terminals, workstations and PCs).
- It should support a hierarchical structure of files, groups and projects.
- It should not add substantially to the overhead of the system.
- It should enable the user to find data on the basis of engineering queries.
- It should be simple to set up and maintain.
- It should be fail safe, i.e. the data should not be corrupted or lost if the index is lost.
- It should enable the system manager to track files and disk usage centrally.
- It should enable backups to be made centrally to safeguard data.
- Applications should be able to make use of the file management facilities offered.

The characteristics of a repository or archive system include the following points.

- It should provide secure storage for data files.
- It should control access to data, preventing unauthorized modification of data.
- It should reflect the working practices of the organization in terms of possible workflows which can be followed.
- It should support a heterogeneous network, with different hardware platforms, operating systems and network protocols.
- It should be usable from many different platforms, including terminals, workstations and PCs.
- It should support data validation procedures, to ensure the consistency and correctness of descriptions and identifiers.
- It should support complex data structures and relationships and so will probably require a relational database.
- It should provide backup, archive and retrieval facilities to a variety of secure media.
- It should be set up and maintained centrally and should have administrator facilities.
- It should be fail safe, providing recovery facilities for the identifiers and descriptive data (metadata).
- It should enable the system manager or administrator to allocate facilities and quotas and monitor system facility usage centrally.
- It should provide secure data storage facilities for applications.

The facilities offered by a data registration and management system should be modular and should include:

- creation of files following the standards laid down by the organization, including the use of specified seed files, units, drawing borders and so on;
- registration of existing data, following the same standards;

- implementation of central data validation standards for identifiers and descriptors;
- central management and administration of data;
- support for complex data structures following the working practices of the company;
- support for projects and sub-projects;
- support for revision or release control;
- support for engineering change order processing.

The facilities offered by a document capture and management system should include:

- conversion of engineering drawings, at least to A0 size;
- conversion of multi-page business documents, A3 and A4, with the ability to change page numbers, etc;
- storage of documents in a standard form;
- ability to view documents in native mode, e.g. raster, word processor and CAD format;
- ability to annotate the documents with text or graphics (redlining);
- optional intelligent character recognition capabilities for text documents;
- ability to easily retrieve documents based on keywords or content of documents (full text retrieval);
- ability to edit raster documents to create a new document or revision.

The facilities offered by a fully customized technical information management system by their nature vary, but should at least encompass some or all of the components outlined above. The basic facilities which should be offered by the customization tools include the following capabilities.

- Prototyping, so that users can provide feedback at an early stage of development.
- Structured development support, so that the customization is maintainable.
- Multi-media support, so that graphics, database, templates, menus, screens, forms and files of various types and formats can be used.
- Code and template management support, so that several people can work together on the customization project.
- Documentation support, so that the customization can be fully documented.
- Ease of use, so that customization can offer high quality results quickly.

If no customization tools are provided, or they are difficult to use and open to use in an unstructured fashion, then any customization will be both inefficient to produce and difficult to maintain.

The points outlined above describe the basic characteristics of the technical information management areas. It should be possible to customize any system to take account of specific customer requirements. In particular, it should be possible to expand the functionality available in a technical information management system to collect, control and manage data from many sources to form an integrated system.

12.6   THE INTEGRAPH TIM PRODUCTS

The Integraph TIM products form an integrated family and are used as the basic data management mechanism for many applications. The products are as follows

- The active data manager, ADM, which provides basic file management functionality over the network.
- The repository/network file manager, I/NFM, which provides repository management functionality; workflow definition and dynamic access control facilities, multi-media support and extensive backup and archive facilities over the network.
- The document management system, DMANDS, which provides management and distribution facilities for raster, vector and ASCII documents, support complex relationships, provides view and redline (annotation) facilities, and provides plotting capabilities.
- The customization toolbox, DB Access, which provides facilities for customization of the TIM products, including access to multiple databases, reporting facilities, graphics capabilities and a high-level scripting language.

There is also a TIM basic customization project, which can be seen as addressing a core requirement in the management of engineering drawings in the drawing registration and management system (DRAMS, produced by Integraph UK), which provides an engineering interface to drawing management, based on the functionality of I/NFM and DB Access. The facilities provided by these products are outlined in the following sections, together with a description of the facilities provided by the DRAMS project, and a fuller description of their capabilities is given in the series of technical overviews available for the TIM products. All products which use a relational database are based on RIS, the relational interface system, ensuring that they are database independent and will operate in exactly the same way for each supported RDBMS on each supported platform.

12.6.1   ADM, the active data manager

ADM is a network file management system which uses metadata to manage files. ADM manages data objects and folders; these are similar in concept to files and directories, but their physical locations are transparent to the user. It will support logical groupings of data, such as projects, and provides functionality to support the creation, movement, copying, deletion, backup, archiving and restoration of data objects and logical groupings of data regardless of where they reside in a network.

ADM will incorporate linkages to I/NFM, so that applications can take advantage of the facilities of the repository, such as secure storage and dynamic access control. For example, data can be retrieved from a repository for modification as a logical group, managed by ADM during modification, and returned to the repository as a new version or revision after modification.

ADM will provide the basic file management capabilities across the network and can be customized to reflect the user's requirements. It can be used as a stand-alone product or can be incorporated into a customized TIM system.

12.6.2   I/NFM, the repository manager

I/NFM is a system for managing files, both current and archived, in a non-homogenous network environment. I/NFM provides a metadata or library system based on a relational database for file management and access, and can be used as a standard platform for data management applications for both Integraph and user-developed applications.

The information about files is organized into catalogs, whose attributes can by defined and modified. Synonyms, default values and validation criteria can be defined for the attributes. Workflows can be defined for groups of data items, specifying the states through which a data item must pass before it can be considered to be complete, and access to data can be restricted dynamically depending on the state the data item has reached, the type of access required (update, review, delete, etc.) and on the class of user requesting access.

The central repository of information holds the physical locations of the files and their access control requirements, and the user can browse through the catalog entries and select a file, which I/NFM will copy from its permanent storage location to the user's working area for access. Only one user can update a file at any time; however, many users can display a file simultaneously (i.e. have read-only access). When a user has completed their modifications, I/NFM returns the file from the work area to its permanent storage location, as a new version of the file.

I/NFM has archive and retrieval functionality so that archived files can be managed in the same way as active files. Files can be archived to tape, magnetic disk or optical disk. Groups of related files can be backed-up for project snapshots and file distribution.

12.6.3   DMANDS, the document manager

DMANDS is an applications product for providing easy access to a central archive of released colour or black and white drawings/documents for the viewing, redlining (mark up), plotting and/or searching/reporting needs of corporate design, manufacturing, maintenance or support personnel. Based on Integraph's common network file management product (I/NFM), DMANDS provides Integraph customers with the ability to implement document distribution with any desired level of access control enforcement.

DMANDS Manager provides a collection of utilities for converting existing documents into the DMANDS native storage formats, bulk loading of document attribute information into the document management database and performing rapid extraction of documents stored on optical disk media.

DMANDS Sponsor enables the Integraph DMANDS View and Redline products for both workstations and PCs to query the central database, select documents for local viewing/redlining/plotting, transfer the document index information and archived files to the local site and then perform view, redline, search and plot operations.

DMANDS View (Workstation/PC) allows users of the DMANDS system to view documents/redlines stored in the central archive. DMANDS View offers the same functionality to PC users that is afforded to Integraph workstation users. Users may view colour raster files, black and white (binary) raster files or ASCII text files. All file types are supported for multiple page operations. Additionally, DMANDS View users may search for the occurrence(s) of text strings within ASCII documents. A maximum of four documents of any type may be viewed simultaneously – each with a maximum of 14 active redline layers.

DMANDS Redline (Workstation/PC) provides users at either MS-DOS compatible PCs or Integraph workstations with all of the functionality of the DMANDS View product. In addition, users of DMANDS Redline may perform non-destructive redline (mark up) operations on documents maintained in the central archive. Each document accessed by DMANDS may have up to a total of 256 redline layers.

12.6.4   DB Access, the customization toolbox

DB Access customization toolbox is Integraph's custom system development/rapid prototyping tool for Integraph workstations. DB Access allows an application developer to rapidly create customized, menu-driven applications which may access or update information stored in one or more databases supported by Integraph's relational interface system (RIS) product (Informix, Oracle, Ingres and DB2 are currently supported).

DB Access includes a powerful report generator which allows the rapid creation of customized reports from databases supported by RIS (all DB Access reports are fully compatible with the RIS Reportwriter and RIS Dataview products also offered by Integraph).

DB Access is capable of creating reports and forms which combine data from multiple databases. In addition, when DB Access is used in conjunction with Microstation, application developers can create systems which allow end-users to access database information via vector graphic element selection or vice versa.

When combined with Integraph's network file management product (I/NFM), the built-in support within DB Access allows application developers to integrate powerful custom system development tools with the core file management environment provided by I/NFM. Finally, when combined with the DMANDS Sponsor product, applications developers using DB Access can create environments whereby users may locate and retrieve database information associated with raster graphics elements or vice versa.

12.6.5   DRAMS, the drawing registry and management system project

DRAMS is a system designed to manage engineering drawings: to control their creation, modification, revision and distribution. DRAMS, however, does not concern itself with the content of the files it manages, so it can also be used as an engineering information manager, managing the files associated with drawings, such as bills of materials, word processor documents and so on, as well as the drawings themselves. DRAMS has been created as a modular system, so that expansion to fulfil future needs is not only possible but simple.

Engineering drawings may be created on a CAD system, or may exist in the form of paper drawings, or in a digital form or, possibly, may need to be created as a CAD modification of an existing drawing held in paper or digital form. Drawings held in digital form may be stored in any one of the many CAD formats in use, or in a standard format such as IGES. Thus, DRAMS provides facilities for the input and use of drawings in a variety of formats.

The DRAMS interface targeted at the CAD user is thus a graphical user interface, running either on a workstation or PC. It also offers linkages to Microstation, so that the engineering user can create and modify the drawings within the framework of DRAMS.

The management of engineering change orders (ECOs) is extremely important to the whole engineering process and DRAMS offer several modules covering revision control, drawing change management, change request management, change order management, proposal drawing management and contractors/client work package management. The Integraph TIM (Technical Information Management) products, as they exist or have been specified, fulfil the characteristics specified above.

The specification of ADM (the active data manager), fulfils the bulk of the nucleus file manager requirements and can be considered to have an appropriate level of functionality for this purpose. Thus, ADM is probably the simplest and most appropriate solution to the nucleus file management problem, catering for both the small system case and the project group/work group case.

I/NFM fulfils all of the requirements of the data archive or repository and is very suitable for this role. Its graphic user interface may need customization and the alpha-numeric interface could probably be improved, but, in general, I/NFM is the appropriate product for this purpose.

DMANDS fulfils the requirements of a document management system. It can manage many types of documents of various sizes and formats from many different sources. It is suited to the management of complex relationships between and within documents and provides many view and redline capabilities. Thus, it is an appropriate product for document management, and, through its links with I/NFM, is particularly suited to manage technical documentation, since it can preserve the relationships between the original designs, drawings, text or documents and their representation in the technical documentation.

The requirements of the data registration and management system are reflected in the specification for the DRAMS system. This work is being carried out in response to customer requirements in several industry areas, including oil, electricity supply, power generating and engineering industries. This development is being carried out using DB Access and I/NFM, with many standard modules to cut down the amount of tailoring required for individual customers. The major customization effort will be in the user interface and the data validation routines.

12.6.6   Integration environment

Hence, it can be seen that Integraph enables unification into a whole by providing a systems environment with a full file compatibility across many major operating systems including UNIX, PC-DOS, Apple Mac and VAX VMS. The Integraph systems environment has a common graphics nucleus with full support for graphics exchange standards and many of the non-standard graphics translators. Integraph has tremendous strength in networking and supports all the industry standards in this important area. Integraph offers a huge range of integrated applications and through a more recent development (relational interface system – RIS) we provide customers with the ability to access any corporate database held on any of the major relational database systems currently on the market. This is our environment and we believe that in its totality it is unique.

12.6.7   An information management cell

Providing an environment that enables integration is obviously essential, but what do we do with it and how do we use it to improve our business?

To answer these questions we need to go back to some basic truths – some fundamentals of business life. We suggest that we manage information in everything we do in our business life. We can also describe a series of activities required to manage the information.

First, there are series of activities which involve access to information – acquire information, store it and then retrieve it. Very often in performing these activities, we simply go through mental processes or use paper. Integraph now enables these activities to be performed electronically in a cost effective manner that only a few years ago was not possible.

Once we have been through the activities involving access to the information, we use it to add our own contribution, our own added value. Frequently this involves manipulation and analysis of information. In a technical environment this is where interactive graphics and data excel and indeed where CAD itself has come from.

One we have added our contribution it is normally desirable, if not necessary, to deliver it somewhere else in the organization, or perhaps to customers, or suppliers, and to do this we need to present it, and subsequently, communicate it.

So if we look at these activities as being a cell involving information management, we are constantly performing activities that involve access to information, use of information and delivery of information. An information management cell could be an individual design engineer, a design department or even the business as a whole.

Technology moves ever forward and now Integraph offers the possibility of electronically linking and automating each of these activities within an information cell, however big or small it is. Perhaps even more importantly Intergraph offers electronic linking and automating one information management cell to another through an integrated systems environment. Now perhaps it is possible to begin to see how the benefits of integration might apply to your business.

12.7 COMMITTED MANAGEMENT

Concurrent engineering may mean different things to different organizations, depending on their methods of conducting business and the products with which they are concerned. But, irrespective of the type of industry, systems to support a concurrent engineering environment must be able to satisfy a number of fundamental requirements, as we have seen.

From a management standpoint, to make concurrent engineering work requires a top-down commitment. It will probably mean a total rethink of current operating practices, beginning with procurement procedures and followed by harmonious and parallel involvement of all design, manufacturing, marketing and service and support groups.

Meanwhile, the computer systems to support such an environment must at least provide the following capabilities.

1. The ability to provide true workflow integration, through a common data format and consistent user interface, irrespective of engineering discipline.
2. The ability to allow designers and engineers from different disciplines and departments to exchange and share information electronically without the need for data translation or re-entry, or to know where the data is physically located.
3. The ability to provide routines to manage responses to engineering change orders, to control who has access to what data, who has authority to amend what, and to provide an audit trail.
4. The ability to manage the entire document and technical information flow, irrespective of whether the data has been created manually or digitally.
5. The ability to protect existing investments in computers and data-bases, and to allow them to be incorporated into the concurrent engineering network.

On the positive side, computer systems to support concurrent engineering do not have to be implemented all at once. Implementation can be gradual – but there must be a committed plan.

12.8   CONCLUSION

In summary, Intergraph is in a position to provide computer-aided engineering (CAE) systems with very advanced system integration through its range of tools designed to provide a fully integrated environment. These tools are able to offer the opportunity to bring technical information out of its technical environment and make it available to everyone in an organization who might need it through an advanced technical information management system.

CHAPTER 13

Strategies for concurrent engineering and sources of further information

C. S. Syan and J. V. Chelsom

13.1 INTRODUCTION

This chapter summarizes the strategic issues in practice of concurrent engineering. Also sources of further information, including published papers, books and suppliers of support tools for CE are listed.

13.2 A PRE-REQUISITE FOR SURVIVAL

Some companies have grown up using new product development processes that incorporate the essential elements of concurrent engineering. Others have recognized the value of the approach and are striving to introduce it. Time is running out for the rest.

The need for CE has been staring Western companies in the face for more than 20 years. As long ago as 1990 the *Harvard Business Review* [1] wrote:

> Three familiar forces explain why product development has become so important. In the last two decades, intense international competition, rapid technical advances and sophisticated, demanding customers have made 'good enough' unsatisfactory in more and more consumer and industrial markets.

This is reinforced by the 'Chief Executives' Attitudes to Technological Innovation in UK Manufacturing Industry' survey which identified time to market as the top business issue of the 1990s [2].

The difficulties of changing have also long been apparent. In 1983, writing about the success of General Motors' quality of working life programme, *Fortune Magazine* [3] carried a prophetic statement:

> The prognosis is less certain for GM's overhaul of product planning and development. GM has to change attitudes and procedures among a broad cross-section of managers. For many of them, success will come this year not because they are doing things differently, but because they are doing the same thing that brought success in the past – selling lots of the bigger cars that make money the old fashioned way. Some managers may draw lessons that stiffen their resistance to change.

Ten years later, GM was still struggling to introduce CE when they were overtaken by the combined pressures of a global market downturn and the successful Japanese transplant operations in their own home market. Their plight has strengthened the resolve of other US companies to press on with their survival and recovery plans – Xerox with 'Leadership Through Quality' [4], IBM with 'Market-Driven Quality', for example. Hewlett Packard still does things the HP way, with great success since the HP way is the CE way.

In Europe, ABB has its 'Customer Focus Program' which led to the T50 programme – aiming to take 50% off the time to do everything, including new product introductions. Nestlé has been working to 'bring research closer to the edge of global competitive strategies' [5].

All these programmes embody the principles of CE.

13.3 INTEGRATION WITH OTHER STRATEGIES

Concurrent engineering not only fits in with these survival and success strategies, it is an integral part of all of them. The fundamentals – the critical success factors – are the same for CE as for TQM, employee involvement, participative management, JIT and lean manufacturing. For most Western companies, a strategy for CE therefore has to be part of a strategy of culture change embracing internal and external relationships.

CE teams need empowerment, just like quality improvement teams. They need the same style of reward and recognition, the same drive for never-ending improvement and the same ability to communicate without fear.

CE teams need to listen, just like customer care teams. They need the same early participation of suppliers as JIT planners. They need training in group problem solving techniques, just like process improvement teams.

All these teams need top management commitment and leadership by example – 'walk the talk'. They can be helped by advances in data processing hardware and software, but these are enhancements, not the essence of CE.

13.4   EXTENSION IN SCALE AND SCOPE

Having been re-discovered by the automotive and aerospace industries, and incorporated from the outset by some consumer electronics companies, CE has spread to almost every manufacturing industry. The concept is also catching on in service industries – even higher education is beginning to consider customer needs and to seek ways of designing these into the product and process. This is partly the result of the pioneer industries pulling their ideas through the supply chain at the same time as they seek to establish CE as standard operating practice throughout their own organizations.

An example of the CE explosion is Ford's development of their latest 'world' car – the Mondeo, replacing the Sierra in Europe and the Tempo/Topaz in North America. Code named 'CDW27' because it is a C/D class car for world markets, it was designed in Europe using product planning and marketing inputs from around the world, and recognizing the manufacturing process implications for the producer plants in both North America and Europe. Suppliers were selected for their global capability and competitiveness, with over 400 of them participating in briefing sessions in 1990 to assist their understanding of the programme, and to enhance their contribution to the component review teams and program module teams. These teams covered the whole vehicle, and the design incorporated engine families that had already been developed using the CE approach – the ZETA range from Europe and the Modular V engine range from the USA.

Another project of ambitious scale and scope is the Boeing 777, which incorporates customer airline representatives in the design teams, as well as key suppliers. Customer inputs cover far more than operational performance requirements – flight crew, cabin crew and maintenance personnel all have a say, and their musts and wants are being identified in time to avoid late changes. For example, advanced CAD enables on-screen simulation of maintenance tasks to check accessibility, without building mock-ups, let alone complete planes. The prize may be aircraft two years earlier in service, and the greater market share and quicker pay-back that brings.

13.5   ROADBLOCKS

Some of the cultural barriers to implementing CE have already been noted. Another enormous obstacle is that organizations aimed at introducing CE are in many cases developed using the old hierarchy and personnel. The needs of the customer – in this case those who are to work using the CE approach – may be sub-ordinate to vested interests, in the false hope that CE processes can be fitted to the new organization and the old personalities. This will not work. A successful strategy to introduce CE has to be based on the same open-minded attitudes as CE itself. Another danger is the reversion to old ways of short-term cost reduction in the face of economic downturn. It is self-evident that a company's survival is a pre-requisite for the introduction of CE, but survival depends on preservation of the quality and timing benefits of CE and JIT as well as their cost reductions. Co-operation rather than confrontation becomes even more vital in declining markets. The partnerships that have been built on the basis of CE are going to be needed even more – not just to survive the strains of the downs of the economic cycle, but also to meet the demands of yet more rapid technological change.

13.6   THE CHALLENGE OF NEW MATERIALS

Another dimension has been added to the task of product and process designers. New materials with new properties offer opportunities for new product features, but demand new manufacturing processes if their full potential is to be realized. Time-based competition is now focused on the span between 'blue sky' research and the market place, and shortening this interval is the new challenge for CE. It is now necessary to simultaneously develop new materials, new products and new processes (Figure 13.1), which calls for yet more changes in management of the new product process (Figure 13.2). The one constant factor is the need for partnership relations between and within companies.

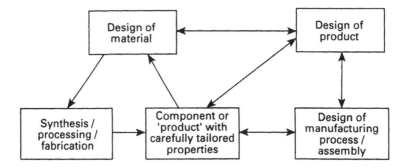

Fig. 13.1   Concurrent design of materials, product and manufacturing: a skeletal representation.

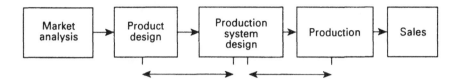

'Normal' relationships, some feedback, simple to manage.

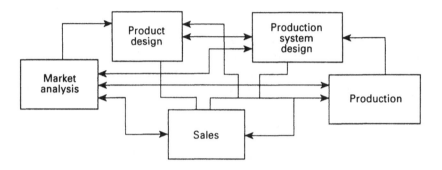

Concurrent engineering, non-sequential and complex to manage.

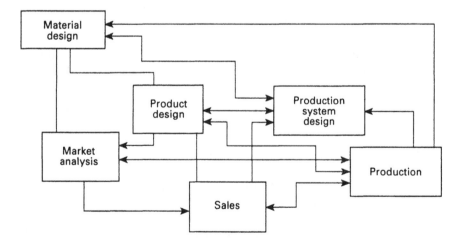

Concurrent design of material, product and process. Very complex management.

Fig. 13.2 Increasing management complexity.

13.7   SOURCES OF FURTHER INFORMATION

This section gives supplementary sources of further useful information for the reader in important aspects of CE and its practice. These sources do not include the references that are already given at the end of most chapters of this book. This list covers most of the topics covered in this book and beyond.

13.7.1   Further reading

American Supplier Institute (1989) *QFD Implementation Manual for 3 day workshop*, ASI Dearborn, MI.

Andreasen, M.M. and Hein, L. (1992) *Integrated Product Development*, IFS Publications Ltd, Bedford, England.

Cohen, L. (1988) Quality Functional Deployment: An application perspective from Digital Equipment Corporation. *National Productivity Review*, **Summer**.

Clover, V.T. and Basley, H.L. (1984) *Business Research Methods*, Grid Publishing Inc. Columbus, Ohio.

Desa, S. and Schmitz, J. (1991) The Development and Implementation of a Comprehensive Concurrent Engineering Method: Theory and Application. Proceedings of the Aerospace Technology Conference and Exposition, Long Beach, California, 1991. *SAE Technical Paper Series*, 37–45.

DeVera, D. (1988) An Automotive Case Study. *Quality Progress*,

Dubensky, R.G. (1992) Simultaneous Engineering of Automotive Systems – Analytical Techniques. Proceedings of the Worldwide Passenger Car Conference and Exposition, Dearborn, Michigan, 1992. *SAE Technical Paper Series*, 131–9.

Ford Motor Co. (1989) *QFD Training Manuals, 8012-210*, Plymouth, MI 48710.

Harrison, F.L. (1992) *Advanced Project Management*, University Press, Cambridge.

Haug, E.J. (ed.) (1990) *Concurrent Engineering of Mechanical Systems*, The American Society of Mechanical Engineers, New York.

Hayes, R. *et al.* (1988) *Dynamic Manufacturing: Creating The Learning Organization*, The Free Press, New York.

Hoinville, G. and Jowell, R. (1978) *Survey Research Practice*, Heinemann Educational Books, London.

Hollins, B. and Pugh, S. (1990) *Successful Product Design: What to do and When*, Butterworths, London.

King, R. (1987) Listening to the Voice of the Customer. *National Productivity Review*, **Summer**.

Kogure, M. and Akao, Y. (1983) Quality Function Deployment and CWQC in Japan. *Quality Progress*, **October**.

Koksal, G. and Smith, Jr., W.A. (1990) Quality Function Deployment: An Application Perspective for Textiles Dyeing and Finishing. *IE Technical Report,* **90-6**, Dept. of Industrial Engineering, North Carolina State University.

Lochner, R.H. and Matar, J.E. (1990) *Designing for Quality*, Chapman & Hall, London.

McElroy, J. (1989) QFD: Building the House of Quality. *Automotive Industries,* **January**.

Monk, Wright, Haber, *Davenport, Information Systems: Improving your Human Computer Interface*, Prentice Hall, NJ, USA.

Moser, C.A. and Kalton, G. (1985) *Survey Methods In Social Investigation*, Gower Publishing Ltd., Hants.

Newman, R.G. (1988) QFD involves buyers/suppliers. *Purchasing World,* **October**.

Proceedings of the 1990 ASME International Computers in Engineering Conference. 217–24.

Pugh, S. (1991) *Total Design – Integrated Methods for Successful Product Engineering*, Addison-Wesley, Wokingham UK.

Reinertsen, D.G. and Smith, P.G. (1991) *Developing Products In Half The Time*, Van Nostrand Reinhold, New York.

Ross, P.J. (1988) The role of Taguchi methods and Design of Experiments in QFD, *Quality Progress,* **June**.

Roy, Ranjit (1990) *A Primer On The Taguchi Methods*, Van Nostrand Reinhold, New York.

Shina, S.G. (1991) *Concurrent Engineering and Design for Manufacture of Electronics Products*, Van Nostrand Reinhold, New York.

Shneiderman, B. (1986) *Designing the User Interface Strategies for Effective Human Computer Interface*, Addison-Wesley, Wokingham, UK.

Singh, K.J. *et al.* (1992) DICE Approach for Reducing Product Development Cycle. Proceedings of the Worldwide Passenger Car Conference and Exposition, Dearborn, Michigan, 1992. *SAE Technical Paper Series*, 141–50.

Slack, N. (1991) *The Manufacturing Advantage*, Mercury Books, London.

Sommerville, I. (1992) *Embedded Systems Software Engineering*, Addison-Wesley, Wokingham, UK.

Stark, John (1992) *Engineering Information Management Systems*, Van Nostrand Reinhold, New York.

Syan, C.S. and Pawar, K.S. (1991) *Sequential or Simultaneous Engineering: A Choice of Strategy*. Transformation of Science and Technology into Productive Power, Proceedings of the 11th International Conference on Production Research, August 1991, China, pp. 1232–6.

Syan, C.S. and Rehal, D.S. (1991) *Design to Market: An Integrative Approach*. Proceedings of the 1st International Conference on Computer Integrated Manufacturing, October 1991, Singapore, pp. 307–10.

Thierauf, R.J. (1986) *Systems Analysis and Design: A Case Study Approach*, Merrill, USA. Distributed by; International Book Distributors, Campus 400, Mayland Avenue, Hemel Hempstead, Herts.

Twigg, D. and Voss, C.A. (1992) *Managing Integration in CAD – CAM And Simultaneous Engineering*, Chapman & Hall, London , UK.

Watkins, K. (1993) *Discrete Event Simulation in C*, McGraw Hill, NY, USA.

Yourdon, (1988) *Modern Structured Analysis*, Prentice Hall, NJ, USA.

13.7.2   Some supporting tools for CE

1. **Statemate** (Software Prototyping Systems)
   I-Logix inc. 22 Third Ave,
   Burlington, MA 01803, USA.
   Telephone 0101 617 272 8090

   I-Logix UK Ltd. 1 Cornbrash Park,
   Bumpers Way, Chippenham, SN14 6RA,
   UK.
   Telephone 0249 446 448

2. **Lucas DFA System**
   Lucas Engineering and Systems Ltd.,
   PO Box 52, Shirley, Solihull,
   West Midlands B90 4JJ, UK.
   Telephone 021 627 3338

3. **BDI – DFMA System**
   Wakefield,
   Rhode Island,
   USA.

4. **CADDS** suite of products (CAD/CAM and Data Management)
   Computervision (UK) Ltd.,
   Argent Court, Sir William Lyons Road,
   Coventry, CV4 7EZ,
   UK.
   Telephone 0203 417718

5. **ICAD** (Knowledge-based CAD/CAM)
   ICAD Engineering Automation Ltd.,
   Viscount Centre II, Milburn Hill Road,
   University of Warwick Science Park,
   Coventry, CV4 7HS,
   UK.
   Telephone 0203 692333

6. **Integraph** suite of products (Integrated CAD/CAM System)
   Integraph (UK) Ltd., Delta Business Park,
   Great Western Way, Swindon, SN5 7XP,
   UK.
   Telephone 0793 619999

7. **QFD/Capture** Software,
   International Technegroup Inc. (ITI),
   5303 DuPont Circle,
   Milford, OH 45150,
   USA.
   Telephone 0101 513 576 3900

8. **American Supplier Institute (ASI)**
   15041 Commerce Drive South,
   Dearborn, MI 48120,
   USA.
   Telephone 0101 313 336 8877 (for courses/training)
             0101 313 271 4200 (publications)

REFERENCES

1. Harvard Business Review (1990) Fourth Quarter Issue.
2. Bone, S. (1992) *Chief Executives Attitudes to Technological Innovation in UK Manufacturing Industry*. P.A. Consultants, London, UK.
3. Fortune Magazine (1993) USA.
4. Friedrich, H.K. (1992) *The Lord Austin Lecture*. Institute of Electrical Engineers, Savoy Place, London, UK.
5. Financial Times, 14th July (1992), *Research comes back to the Nest*. London, UK.

Index

Page numbers appearing in **bold** refer to figures and page numbers appearing in *italic* refer to tables.

CPSIA information can be obtained at www.ICGtesting.com
Printed in the USA
LVOW011248280413

331269LV00004B/98/P